T0140915

Development and evaluation of iodine biofortification strategies for vegetables

by Dipl. Ing. (FH) Patrick G. Lawson
[Registration no. 957853]

Dissertation submitted to the Department of Cultural Studies and Geosciences [FB 02] of the University of Osnabrück in candidacy for the degree of Doctor of Natural Sciences [Dr. rer. nat.]

April 2014

Dean: Prof. Dr. habil. H. Koriath
1st Examiner: Prof. Dr. habil. H. Meuser
2nd Examiner: Prof. Dr. habil. J. W. Härtling
3rd Examiner: Prof. Dr. D. Daum

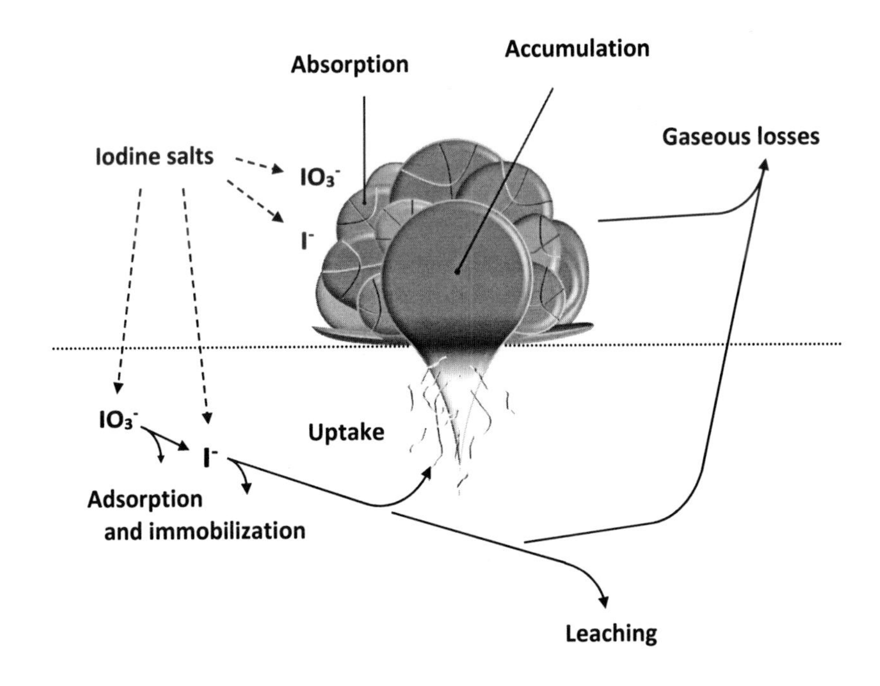

This is a complete printout of the dissertation entitled "Development and evaluation of iodine biofortification strategies for vegetables" submitted by Patrick Grant Lawson to the Department of Cultural Studies and Geosciences of the University of Osnabrück in candidacy for the degree of Doctor of Natural Sciences.

Submitted 28 April 2014 and accepted 4 July 2014. Date of public disputation: 14 July 2014. Members of the doctoral committee: Prof. Dr. habil. H. Koriath, Prof. Dr. habil. H. Meuser, Prof. Dr. habil. J. W. Härtling, Prof. Dr. D. Daum, Prof. Dr. habil. G. Broll and A. Stele.

This dissertation was realized in cooperation with the University of Osnabrück (Universität Osnabrück) and the University of Applied Sciences, Osnabrück (Hochschule Osnabrück), Germany.

Osnabrück, August 2014

Bibliographic information published by the Deutsche Nationalbibliothek

The Deutsche Nationalbibliothek lists this publication in the Deutsche Nationalbibliografie; detailed bibliographic data are available in the Internet at http://dnb.d-nb.de

ISBN 978-3-8325-3779-1

Logos Verlag Berlin GmbH
Comeniushof • Gubener Str. 47
10243 Berlin
Tel.: +49 (0)30 42 85 10 90
Fax: +49 (0)30 42 85 10 92
Internet: http://www.logos-verlag.de

Contents

List of figures

List of tables

List of abbreviations and acronyms

µg	Microgram
$\mu g\ I\ (100\ g\ FM)^{-1}$	Microgram of iodine per one hundred gram of fresh matter
$\mu g\ I\ d^{-1}$	Microgram of iodine per day
µm	Micrometer
$\mu S\ cm^{-1}$	Electrical conductivity in micro-Siemens per centimeter
^{125}I	Iodine-125 isotope
^{129}I	Iodine-129 isotope
^{131}I	Iodine-131 isotope
AES	Atomic emission spectrometry
Al	Aluminum
AM	Antemeridian
ANOVA	Analysis of variance
As_2O_3	Arsenic (III) oxide
atm	Standard atmosphere
B	Boron
$Ba(OH)_2$	Barium hydroxide
$BaCO_3$	Barium carbonate
BrO_3^-	Bromine trioxide
Ca	Calcium
$CaCl_2$	Calcium chloride
CAL	Calcium-Acetate-Lactate extraction solution
$Ce(CO_3)_5^{6-}$	Cerium (IV) carbonate
$Ce(CO_3)_6^{8}$	Cerium (IV) carbonate
$Ce(SO_4)_2$	Cerium (IV) sulfate
CH_2ClI	Chloroiodomethane
CH_2I_2	Diiodomethane
CH_3	Methyl group
CH_3Br	Methyl bromide
CH_3I	Methyl iodide
CHI_3	Iodoform
Cl^-	Chloride
ClO_4^-	Perchlorate
cm	Centimeter
CN^-	Cyanide
CO_2	Carbon dioxide
CO_3^{2-}	Carbonate

C_{org}	Organic carbon
Cu	Copper
CV; cv.	Coefficient of variation; cultivar
d	Days
df	Degrees of freedom
DIN	Deutsches Institut für Normung (German Institute for Standardization)
DM	Dry matter
DPPH	2,2-Diphenyl-1-picrylhydrazyl radical
EC	Emulsifiable concentrate; electrical conductivity
EFSA	European Food Safety Authority
E_h	Redox potential
EU	European Union
Fe	Iron
Fe^{2+}	Ferrous iron; iron (II)
FIA	Flow Injection Analysis
FRAP	Ferric reducing activity of plasma
g	Gram; G-force
$g\ cm^{-3}$	Volumetric mass density; grams per cubic centimeter
GH	Greenhouse
h	Hour
$h \cdot v$	Photon energy; Planck constant times light frequency
H_2O_2	Hydrogen peroxide
H_2SO_4	Sulfuric acid
ha	Hectare
HCl	Hydrochloric acid
Hg^+	Mercury
HI	Hydrogen iodide
HIO	Hypoiodous acid
I^-	Iodide
I	Iodine
I_2	Elemental iodine
ICP-MS	Inductively coupled plasma mass spectrometry
IDD	Iodine deficiency disorders
IFP	Iodine fixation potential; iodine fertilizer prototype
IO_3^-	Iodate
ISO	International Organization for Standardization
K	Potassium; Kelvin
K_2CO_3	Potassium carbonate
kg	Kilogram
KI	Potassium iodide
KIO_3	Potassium iodate
klx	Kilolux

KOH	Potassium hydroxide
L; L.	Liter; taxonomical classification according to Linnaeus
M	Mole
m^2	Square meter
MeBr	Methyl bromide
MeI	Methyl iodide
Mg	Magnesia
mg	Milligram
mL	Milliliter
mm	Millimeter
mM	Millimole
Mn	Manganese
MnO_4^-	Permanganate
N; n	Nitrogen; numerous
Na_2O_2	Sodium peroxide
$NaCl_2$	Sodium chloride; table salt
NaOH	Sodium hydroxide
NH_4	Ammonia
nm	Nanometer
N_{min}	Mineral nitrogen
NO_3^-	Nitrate
NS	Not significant
O_3	Ozone
P; p	Phosphorous; probability
POD	Point of deliquescence
QE	Quercitin equivalents
R^2	Coefficient of regression
RDI	Recommended daily intake
RH	Relative humidity
Se	Selenium
SEM	Scanning Electron Microscopy
SNC^-	Thiocynates
SO_4^{2-}	Sulfate
TEAC	Trolox equivalent antioxidative capacity
TMAH	Tetramethylammonium hydroxide
v/v	Percent concentration as volume of solute per volume of solution
W	Watt
w/v	Percent concentration as weight of solute per volume of solution
WHO	World Health Organization
X	Times; multiplied by; value
Zn	Zinc

1 General introduction

Iodine is an essential element for human life which must be regularly ingested along with other mineral and organic micronutrients (e.g. iron, zinc, vitamin A and folate) to guarantee a proper development and function of the human organism. Almost one-third of the population worldwide is currently suffering from micronutrient malnutrition (FAO 2013). The so-called "hidden hunger", i.e. not the lack of calories but the lack of one or a number of the essential elements for human life, is the cause of many preventable diseases in the world. Remarkably, micronutrient deficiencies are an issue even in well-developed industrialized countries. The fortification of food with micronutrients during processing and preparation has brought a certain degree of alleviation, but has not always been very successful (HIRSCHI 2009; WHITE AND BROADLEY 2009).

The biofortification of crops is regarded as a powerful method for tackling micronutrient deficiency and has been defined as: "The process of increasing the bioavailable concentrations of an element in edible portions of plants by agronomic intervention or genetic selection" (WHITE AND BROADLEY 2005). The term "biofortification" was originally coined in 2001 by Steve Beeb, head of the current HarvestPlus® program, and it was intended to improve the nutritional value of crops by conventional genetic selection (MORGAN 2013). Therefore, particularly mineral nutrient efficient genotypes among the cultivars of rice, maize, wheat, cassava, pearl millet, bean and sweet potato, have been selected as being the key staple foods in Asia and Africa (ZHAO AND SHEWRY 2011). Nevertheless, in several soils the occurrence and phytoavailability of certain micronutrients is limited and a simple and inexpensive method for enhancing their content in edible plant parts is the application of fertilizers. Especially the elements zinc, selenium and iodine are promising successful agronomical biofortification possibilities, since the genetic improvement in crop mineral uptake will probably have very limited effects when these elements are not additionally fertilized (DAI ET AL. 2004a; HARTIKAINEN 2005; ALLOWAY 2008; CAKMAK 2008). On the other hand, factors such as the application method, soil properties and composition, mineral mobility and accumulation in the plant, may complicate the exploitation of the agronomical biofortification approach (ZHU ET AL. 2007) and must be therefore further investigated.

Improved crop production technologies are the sum of a developmental process which involves a series of elimination procedures. At the base of this development, a broad range of available approaches are evaluated and the product is a new, superior technology which can be recommended for commercial use (GOMEZ AND GOMEZ 1984).

In this dissertation, the agronomical biofortification of crops was used to further rectify the issue of human iodine deficiency in Germany and other countries. The main objective was to develop a feasible iodine biofortification strategy mainly for field-grown vegetable crops. Attempts were made to:

- Identify suitable vegetable species for iodine biofortification and determine their specific fertilization levels. Therefore a selection of economically important crops was compared in their accumulation behavior.

- Ascertain the most appropriate iodine fertilization form. Pure KI and KIO_3 salts were compared.

- Assess the most efficient iodine application technique by comparison of soil and foliar application approaches.

- Determine cultivation and environmental factors affecting iodine accumulation.

- Test an iodine fertilizer prototype and the miscibility of different agrochemicals with iodine under practical conditions.

2 State of the art in science and technology

2.1 Iodine

Iodine was accidentally discovered by the French chemist Bernard Courtois in 1811. The largely government-controlled saltpeter trade of that time forced the manufacturers (*salpětrier*) to look for a substitute for wood ash. Kelp turned out to be a good alternative. During the inspection of his equipment, Courtois noticed a violet vapor (Figure 2.1 A) emerging from some corroded copper vessels and it appeared again after adding sulfuric acid to his seaweed ashes as a result of the thermal influence. The name iodine, which is derived from the Greek *iodos,* meaning violet, was proposed by Joseph Louis Gay-Lussac in 1813 who researched the newly discovered substance at the same time as Humphry Davy (SWAIN 2005).

Iodine is symbolized by the letter "I" and, with an atomic weight of 126.90447 g mol^{-1}, is the heaviest element commonly needed by living organisms. Iodine is positioned 53rd in the periodic table of elements and belongs to the halogen group, the so-called salt formers, which are characterized by their high reactivity with most other elements and compounds. It has an electronegativity (EN) of 2.4 and the oxidation state is mostly -I, but +I, +III, +V and +IIV can occur as oxoacids or, generally with K or Na as counter ions, as salts. Iodine is hardly soluble in water (apart from the chemical bonds HI and KI), but exhibits a high solubility in organic solvents like alcohol, ether or acetone (WHITTEN ET AL. 2007; LATSCHA AND MUTZ 2011).

Iodine is a rare element with no comparable substitutes for many technical and medicinal applications. In the past, it was widely used as a disinfectant for cuts (Figure 2.1 C) and to prevent goiter in a variety of dosage forms (Figure 2.1 D). Nowadays, its most important applications are found in contrast media (Figure 2.1 B), iodophors biocides, pharmaceutical products, synthetic fabric treatments (nylon), human and animal nutrition, liquid crystal displays (polarizing film) and as a catalyst. The current market for crude iodine amounts to > 30,000 metric tons p.a. and it has shown rising prices (from 24 US$ kg^{-1} in 2008 to 41 US$ kg^{-1} in 2012) and strongly growing production rates over the last years (3.5 - 4.0 %). The most important, economically-exploitable source for crude iodine is the caliche ore in Chile (59 % of the worldwide production), were iodine is a by-product of the Chilean nitrate production (POLYAK 2013; SQM 2013).

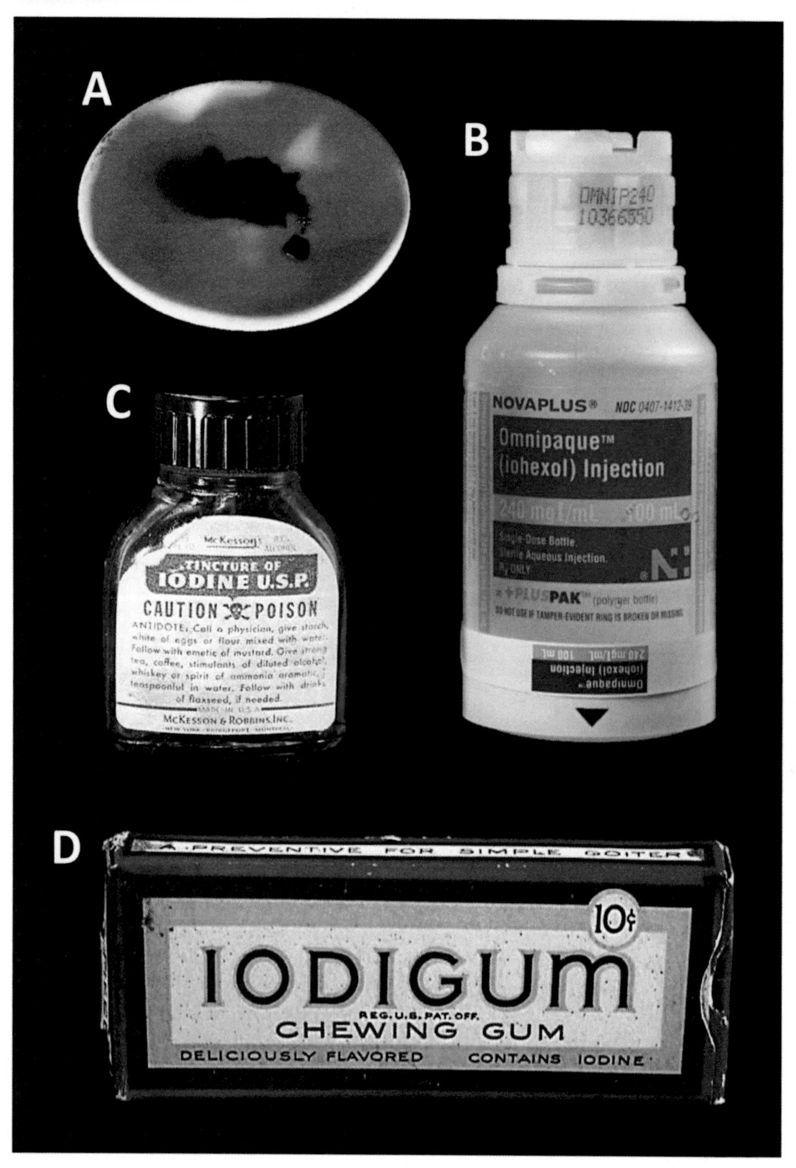

Figure 2.1 A = violet vapor of elemental iodine heated on a white dish (object illuminated from below). B = Iohexol®, an iodine-containing contrast agent used in imaging the heart in CT-scans. C = tincture of iodine for the disinfection of cuts. D = Iodigum®, an iodine-containing chewing gum to prevent goiter in the 1950s years. Pictures by GRAY (2010); reproduced with permission of the author

2.2 Iodine in humans

Iodine plays a vital role in the human organism and must be regularly supplied in a sufficient quantity to ensure the proper function of fundamental physiological processes. It is required for the synthesis of the thyroidal hormones triiodothyronine (T3) and thyroxin (T4). These hormones contribute to the regulation of important functions in the mammal organism such as metabolic processes, growth and brain development. They influence the carbohydrate, fat, protein and mineral metabolism as well as the central nervous system, the neuromuscular transmission and the musculature (PFANNENSTIEL AND SALLER 1991, HINZE AND KÖBBELING 1992).

A low dietary supply of iodine is the main factor responsible for iodine deficiency and a chronic shortage can lead to a multitude of disorders (GROẞKLAUS 1994). An overview of the iodine deficiency disorders (IDD) spectrum across the lifespan is given in Table 2.1. Cretinism is the most severe expression of IDD which can occur either with neurological specificity showing mental defects, hearing loss, lalopathy and ataxia, or with myxedematous specificity in a hypothyroid metabolic status or both (HETZEL ET AL. 1988).

Table 2.1 The IDD-spectrum across the lifespan (adapted from WHO 2004)

Stage of life	Disorder
Fetus	Abortions; stillbirths; congenital anomalies; increased perinatal mortality; endemic cretinism; deaf mutism
Neonate	Neonatal goiter; neonatal hypothyroidism; endemic mental retardation
Child and adolescent	Goiter; (subclinical) hypothyroidism; (subclinical) hyperthyroidism; impaired mental function; retarded physical development
Adult	Goiter, with its complications; hypothyroidism; impaired mental function; spontaneous hyperthyroidism in the elderly; iodine-induced hyperthyroidism
All ages	Increased susceptibility of the thyroid gland to nuclear radiation

The most apparent clinical manifestation of iodine deficiency is the compensatory extension of the thyroid gland, known as goiter or struma (Figure 2.2 shows two different manifestations of goiter). The appearance of this symptom represents the maladaptation of the thyroid gland to iodine scarcity and results from excessive thyroid stimulation in order to exploit physical iodine reserves (DELANGE ET AL. 1985; DUMONT ET AL. 1995).

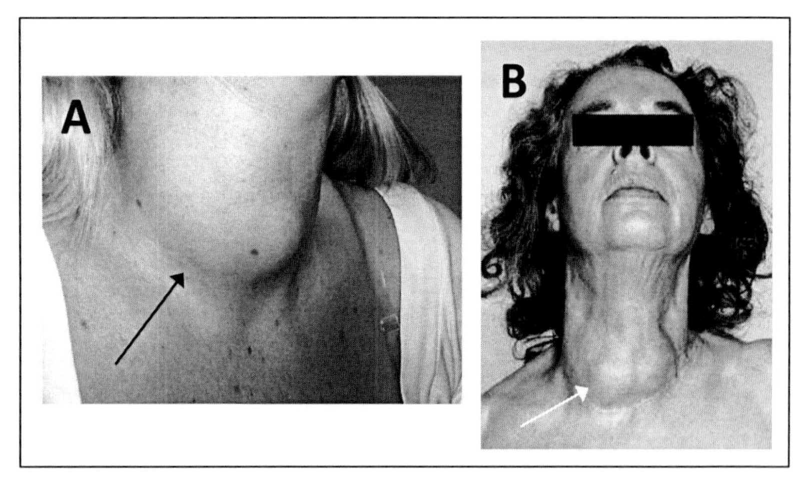

Figure 2.2 A = woman with a typical Derbyshire neck (black arrow), graded as a class II goiter (ENACADEMIC 2014). B = woman with a large multinodular goiter (white arrow; WILLACY 2012)

Furthermore, an insufficient iodine supply can cause various adverse physical effects ranging from mild symptoms such as tiredness and a lack of concentration to severe symptoms such as irreversible growth and development disorders among children and a higher risk of developing several cancer types among adults (BGVV 2001; GOŁKOWSKI ET AL. 2007; KROHN AND PASCHKE 2009). In many developing countries iodine deficiency inhibits children and adults from exploiting their full intellectual potential; a lowered intellectual capacity of about 10-15 IQ-points has been estimated (UNICEF 2004). The iodine deficiency issue is not restricted to the third world countries. Even in the industrialized states of Europe almost 44 % of school-age children and 44.2 % of the general population have an inadequate iodine input (Table 2.2). Worldwide almost 1.9 billion people have an insufficient iodine intake and

740 million people are actually suffering from IDD. Although the iodine nutrition has shown a slight improvement over the last decade, the global progress status is slowing and the intervention programs need to be extended (WHO 2004; ANDERRSON ET AL. 2012).

Table 2.2 Proportion of population and number of individuals with insufficient iodine intake in school-age children (6–12 years) and in the general population (ANDERRSON ET AL. 2012)

WHO region	Insufficient iodine intake (= urinary iodine < 100 µg L^{-1})			
	School-age children		General population	
	Total numbers (Mio.)	Proportion (%)	Total numbers (Mio.)	Proportion (%)
Africa	57.9	39.3	321.1	40.0
Americas	14.6	13.7	125.7	13.7
Southeast Asia	76.0	31.8	541.3	31.6
Europe	30.5	43.9	393.3	44.2
Eastern Mediterranean	30.7	38.6	199.2	37.4
Western Pacific	31.2	18.6	300.8	17.3
Total	**240.9**	**29.8**	**1881.2**	**28.5**

In particular, the western European countries are affected by iodine scarcity, and IDD in the form of goiter still exists. According to DELANGE (2002), the European states of Belgium, Bosnia, Denmark, France, Hungary, Ireland, Italy, Luxembourg, Portugal, Romania, Slovenia, Spain and Turkey are affected by iodine insufficiency. In Germany, the amelioration of the iodine supply is still proceeding towards the optimum; about 30 % of the population still has a mild to moderate deficiency since not every German region and not every phase of life (e.g. pregnancy and lactation) is provided with enough iodine (MENG AND SCRIBA 2002). More than 40 % of the children and adolescents in Germany are insufficiently supplied with iodine (THAMM 2007). Recently, REMER ET AL. (2012) reported a downward trend in the iodine uptake of the German school children because of dietary changes. Screenings on 100,000 adult people in Germany showed that every second person over 45 years of age is suffering from a distended thyroid gland or the formation of glandular knots. Women are significantly more affected than men and most of these people were not aware of their illness. Every year, more than 120,000 medical thyroid operations are carried out in Germany and the direct costs for the treatment of thyroid diseases amount to 1.6 billion € a^{-1} (SCHUMM-DRÄGER AND FELDKAMP 2007; SCHÜCKING AND RÖHL 2010; STATISTISCHES BUNDESAMT 2011).

Incrementing iodine intake through supplementation or food fortification are the recommended vehicles for controlling IDD (WHO 2004) and the use of iodized table salt in households and in the food processing industry has improved the iodine supply in Germany (SCRIBA ET AL. 2007). On the other hand, the excessive consumption of table salt is regarded as a main source for a number of diseases in industrialized nations. Coronary heart diseases, stroke and myocardial infarction and their physical and financial costs could be averted significantly by reducing table salt consumption (ASARIA ET AL. 2007; BIBBINS-DOMINGO 2010). In Germany, there is an urgent need for action. Current table salt consumption (up to > 10 g d^{-1}) is causing several health problems ranging from hypertension in adults to hypotension in children and adolescents. Thus, a reduction of the salt consumption down to 3.5 - 6.0 g $NaCl_2$ d^{-1} is strongly advised (KNORPP AND KROKE 2010; BFR 2012). Furthermore, WENG ET AL. (2014) demonstrated that iodized table salt added during cooking was not as effective as the use of iodine-rich vegetables (biofortified celery) and iodine-free salt. In fact, the majority of the iodine in the iodized table salt was lost within 5 minutes of cooking. In comparison, the biofortified celery retained 86 % of the initial iodine content after cooking for 5 minutes in the same manner. Even after cooking at 100 °C for 30 minutes about 60 % of the initial iodine was preserved in the edible plant parts.

The effect of antinutrients, i.e. substances that hinder the absorption of essential elements, may further reduce the human uptake of iodine. There are goitrogenic substances that, ingested via the food chain, inhibit the assimilation of iodine by the sodium iodide symporter (NIS): Thiocynates (SNC^-) are secondary metabolites and occur naturally in the genus *Brassica*, e.g. in white cabbage, broccoli and kohlrabi. Nitrate (NO_3^-) is widely applied as a mineral nitrogen fertilizer and is abundant in green leafy vegetables, in anthropogenically polluted groundwater and, added as a preservative, in cured meat. Perchlorate (ClO_4^-) is ubiquitously present in the environment as a result of the anthropogenic influence. It is found in rocket propellant and fireworks or is a byproduct of disinfectants like chlorate and hypochlorite salts (TONACCHERA ET AL. 2004; WHITE AND BROADLEY 2005; PARKER 2009; VOOGT AND

JACKSON 2010). Humans obtain iodine mainly by the ingestion of foodstuffs. The iodine concentration in food varies widely and depends on naturally occurring seasonal variations and fluctuations in the soil and climate conditions as well as the geographical region. The use of iodine additives strongly influences the iodine levels in food. The utilization of iodized feedstuffs for dairy cattle and iodophors for udder disinfection enhances the iodine concentration in milk and dairy products, which currently provide 37 % of the daily iodine uptake of the German population (Figure 2.3).

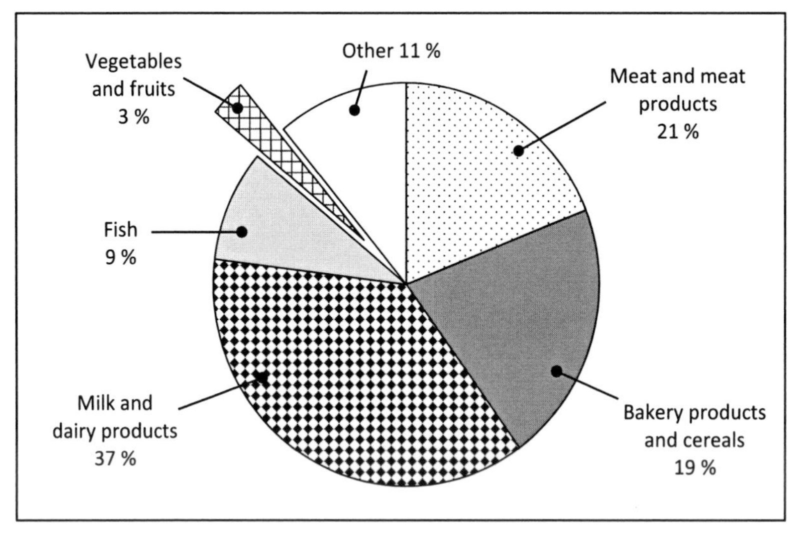

Figure 2.3 Sources of the dietary iodine intake of the German population (GROßKLAUS AND JAHREIS 2004)

Meat and meat products supply 21 % and bakery products and cereals contribute with a share of 19 % to the total intake, mainly because of the utilization of iodized table salt in the food processing. Although products of marine origin are naturally rich in iodine, fish currently only makes up 9 % of the total because of the dietary habits of the German population. Vegetables and fruits only contribute a share of 3 % to the total daily iodine uptake because of their rather low natural iodine levels in the range of 0.5 - 10 µg I $(100\ g\ FM)^{-1}$ (SOUCI ET AL. 2000; GROßKLAUS AND JAHREIS 2004; GRIMMINGER 2005; FUGE 2007; HAMPEL ET AL. 2009). Hence, vegetarians, especially vegans, are at risk of iodine malnutrition if no additional measures for iodine supplementation are taken (REMER ET AL. 1999; KRAJČOVIČOVÁ-KUDLÁČKOVÁ ET AL. 2003). On the other hand, a number of diseases, such as rheumatoid arthritis, stroke, diabetes, coronary heart diseases, several cancer types and Alzheimer´s disease have been correlated to a high consumption of meat and meat products (SONG ET AL. 2004; CHAO ET AL. 2005; TAPPEL 2007; MICHA ET AL. 2010).

In this respect, the need for a well-balanced diet, i.e. low in meat and table salt and rich in vegetables and fruits and therefore low in iodine, emphasizes the importance of an alternative or complementary iodine source, especially exploiting the potential of the biofortification approach (SINGH ET AL. 2003; O'KEEFE AND CORDAIN 2004; TONACCHERA ET AL. 2013).

The recommendations for a sufficient daily supply of adolescents and adults with iodine by several institutions and organizations range between 150 and 200 µg I d^{-1}, but are subject to variations, amongst others, because of body weights being the reference points for these calculations. However, adolescents and adults should ingest 2 µg of iodine per kg of bodyweight daily for a full supply. Pregnant and lactating women have higher iodine needs (Table 2.3). According to the latest nationwide iodine monitoring, the mean iodine intake of the German population was recorded to be 119 µg I d^{-1} and lays, therefore, approximately 30 % under the actual iodine requirements of an adult person (MANZ ET AL. 1998; ANKE AND ARNOLD 2008; ARBEITSKREIS JODMANGEL 2009, 2013a).

Table 2.3 Recommended and actual daily iodine intake showing the resulting nutritional gap. Adapted from ARBEITSKREIS JODMANGEL (2013b)

Phase of life	Recommended intake [µg I d^{-1}]	Actual intake [µg I d^{-1}]	Nutritional gap [µg I d^{-1}]
Children (1 - 9 years of age)	100 - 140	60 - 100	40
Adolescents and adults	180 - 200	120	60 - 80
Pregnant and lactating women	230 - 260	110 - 125	120 - 135

Considering the nutritional iodine gaps indicated in Table 2.3 and an average portion of eaten vegetables, amounting to roughly 80 g of fresh weight per individual per day, a projected iodine concentration of 50 - 100 µg I (100 g FM)$^{-1}$ in biofortified vegetables would, to a large extent, cover the dietary iodine deficiency present in Germany. For higher iodine requirements the vegetable portions could be increased as needed. Thus, a target range of 50 - 100 µg I (100 g FM)$^{-1}$ seems to be a reasonable concentration level for biofortified vegetables that could satisfy the majority of the iodine needs and the dietary habits without running the risk of exceeding the upper tolerable intake level (UL) for iodine.

The UL has been defined by the Scientific Committee on Food as the maximum level of the (continuous) total daily intake of a nutrient, from all sources, considered to be improbable to affect the health of almost all individuals in the general population including sensitive individuals, excepting sub-populations genetically prone to certain diseases (EFSA 2006). The UL can result either from the no observed adverse effect level (NOAEL) minus uncertainty factors or the lowest observed adverse effect level (LOAEL) minus uncertainty factors (Verkaik-Kloosterman et al. 2012). Thus, a large difference in setting threshold levels among the international expert committees and authorities exists and there is no unanimity about the iodine level that should not be exceeded. Richardson (2007) and the IOM (2006) indicate a UL of 940 and 1100 µg I d^{-1}, respectively. The EU Scientific Committee on Food and the European Food Safety Authority define the UL for iodine at 600 µg d^{-1} (SCF 2002; EFSA 2006) whereas the D-A-CH (2000) has set the UL at 500 µg I d^{-1}.

Through an iodine supplemented diet, i.e. with fortified food of animal origin and with a regular utilization of iodized salt, the recommended UL of 500 µg I d^{-1} is not likely to be exceeded and the iodine uptake of the German population can be classified as safe (BfR 2004). In addition, people with a normal thyroid gland function usually tolerate the uptake of large iodine amounts without adverse manifestations (Braverman 1994). Japanese people ingest occasionally amounts of 10,000 – 80,000 µg I d^{-1} through a very rich diet of sea food and algae and this high iodine uptake has been reported to be without unfavorable effects (FAO/WHO 2004). A healthy thyroid can thus adapt to an iodine excess by means of autoregulation mechanisms. On the other hand, people living in iodine deficient regions with endemic goiter and who are 40 years of age or older, may develop iodine induced hyperthyroidism and other diseases if large amounts of iodine (doses of several milligrams per day) are ingested over a short period of time; hyperthyroidism, hypothyroidism or thyrotoxicosis have also been reported after the injection or ingestion of iodine containing drugs such as contrast media for radiology or amiodarone (75,000 µg I $tablet^{-1}$) for the long-term treatment of cardiac arrhythmias (STEIDLE 1989; ROTI AND DEGLI UBERTI 2001). As a consequence of the iodine fortification of food in Germany, criticism has been levied from iodine susceptible individuals suffering either from iodine hypersensitivity, Hashimoto's or Graves' disease, hyperthyroidism or thyrotoxicosis. The reproaches were invalidated by many studies that proved no outstanding health risks from the use of iodized salt and feedstuffs. Merely the consumption of highly concentrated iodine sources, such as dried seaweed products [20,000 – 500,000 µg I $(kg\ DM)^{-1}$] over a long period of time has been classified as potentially hazardous, especially in the elderly with unknown thyroid autonomy, and should therefore be avoided (GÄRTNER 2000, GROßKLAUS AND JAHREIS 2004; BFR 2007; JOHNER ET AL. 2011).

2.3 Iodine in the ecosystem

The potential roles of iodine in living organisms have been localized into a self-defense mechanism and a direct antimicrobial effect in marine algae; herbivores and higher order predators probably use iodine as a surrogate sensor, i.e. as an indicator for the overall availability of ecosystem resources. As a consequence, the modulation of growth, reproduction, metabolic rate and lifespan by the thyroid hormone corresponds to the availability of iodine (YUN AND DOUX 2009).

The geochemistry of iodine has been thoroughly studied and the interactions of numerous bio-physico-chemical properties were ascertained to be heavily influencing the iodine fixation potential (IFP) of soils and, consequently, the iodine influx to humans. Iodine is a rare element in the earth's crust with a total estimated mass of 3.4×10^{15} kg. The mean concentration of iodine in soils worldwide has been assessed at 5.1 mg I $(kg\ soil)^{-1}$. Taking the skewed nature of the collected data into account, the geometric mean, amounting to 3.0 mg I $(kg\ soil)^{-1}$, seems to be a more appropriate value. Sea water contains approximately 50 - 60 µg I L^{-1} and is the major deposit of iodine solutes. The theories of the provenience of iodine have been subject to a divided opinion: The weathering out of parent material and the subsequent run-off into the sea is a possible explanation. On the other hand, iodine may have always been in the oceans and continually accumulates in the upper layer of the earth crust due to the occurring iodine recirculation via the atmosphere. However, the distribution of iodine in the ecosystem is a rather complex cycle (Figure 2.4) which takes part in all states of aggregation and even influences the ozone layer in the stratosphere (WHITEHEAD 1984; JOHANSON 2000; JOHNSON 2003).

In seas and oceans, iodine is predominantly present as iodide and iodate and both iodine forms are subject to several redox changes and processes which may be spontaneous or due to the activity of a number of different bacteria, phytoplankton and algae. Spontaneous processes are the hydrolysis of I_2 to form HIO and the disproportionation of HIO to form IO_3^- in an alkaline environment (Figure 2.4 bottom right). The marine accumulation of iodine may then follow by means of the incorporation activity of algae and bacteria (e.g. *Arenibacter* sp. C-21).

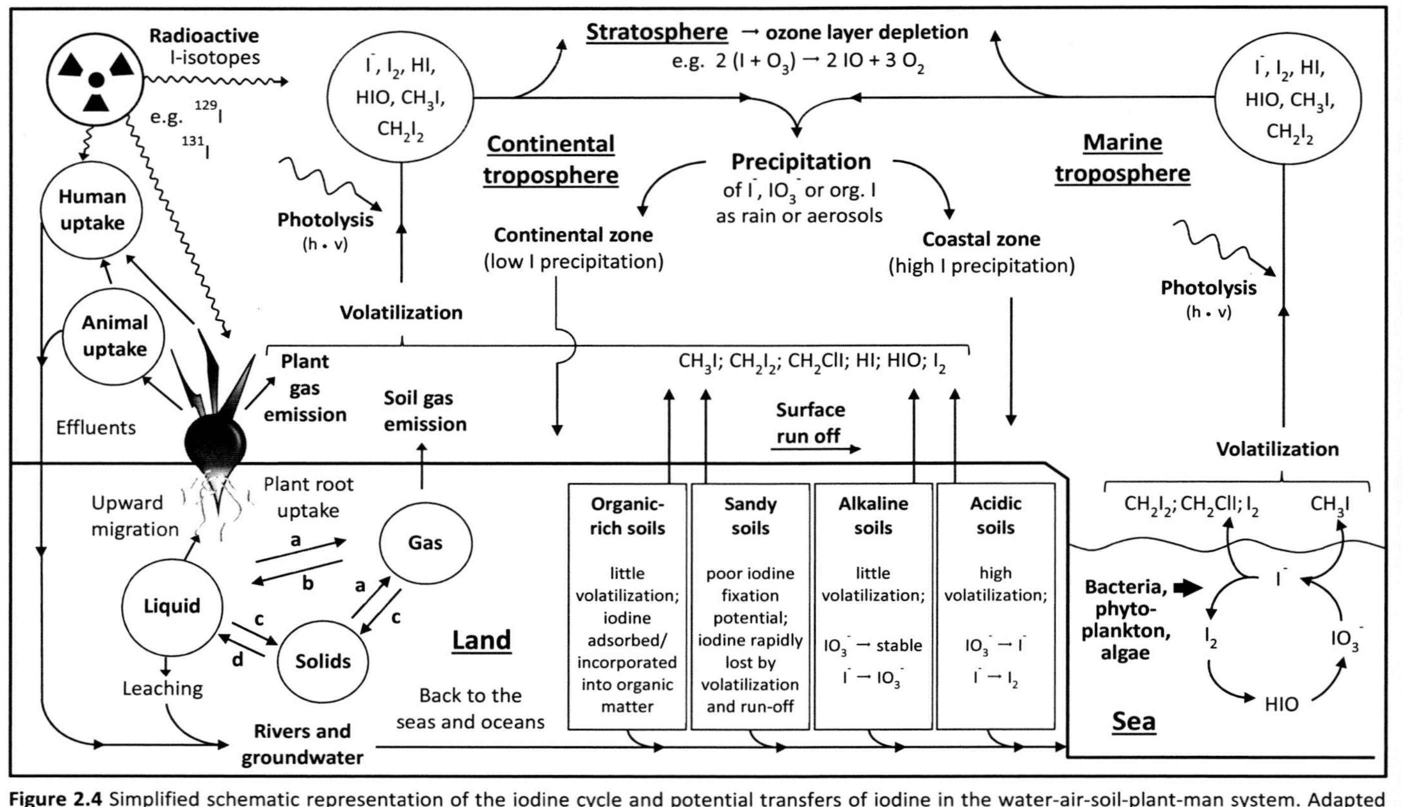

Figure 2.4 Simplified schematic representation of the iodine cycle and potential transfers of iodine in the water-air-soil-plant-man system. Adapted from FUGE (2005), AMACHI (2008) and ASHWORTH (2009). Possible iodine transfer between the soil phases: (a) volatilization; (b) dissolution; (c) adsorption and immobilization; (d) dissolution, desorption and mineralization. Further explanations in the text

The subsequent reduction of IO_3^- to I^- can result from phytoplankton as well as nitrate and iron reducing bacteria (e.g. *Pseudomonas* sp. SCT). Successively, algae, phytoplankton and various bacteria contribute to the methylation of iodine, i.e. the formation of gaseous methyl iodide (CH_3I) by transferring a methyl group (CH_3) to I^-. Additionally, the bacteria *Roseovarius tolerans* and *Rhodothalassium salexigens* may not only produce volatile organic iodine compounds such as diiodomethane (CH_2I_2), chloroiodomethane (CH_2ClI) and iodoform (CHI_3), but also oxidize I^- to I_2 (FUSE ET AL. 2003; AMACHI 2008).

In the atmosphere, volatile organic iodine compounds will partly be converted to inorganic iodine forms through the effects of solar radiation, i.e. through photolysis (e.g. $CH_3I + h \cdot v \rightarrow CH_3 + I$), and the marine troposphere will enrich in different iodine forms and compounds (e.g. I^-, I_2, HI, HIO, CH_3I, CH_2I_2). Some of these may contribute, once arrived in the stratosphere together with other volatile halocarbons (e.g. CH_3Br), to the depletion of the ozone layer (e.g. $2\ I + 2\ O_3 \rightarrow 2\ IO + 3\ O_2$). However, the majority of the inorganic and organic iodine forms and compounds may precipitate down to the land surface as rain or aerosol at concentrations in the range of 0.5 - 5 µg I L^{-1} (DAVIS ET AL. 1996; TRUESDALE AND JONES 1996; KEPPLER ET AL. 2000; FUGE 2005; NEAL ET AL. 2007; AMACHI 2008; HOU ET AL. 2009).

Once precipitated onto the upper soil layer, iodine will again be involved in several processes shifting in emphasis in dependency of the soil type and its bio-physico-chemical properties, namely the organic matter content, redox potential (E_h), pH value, soil texture, iron and aluminum oxides, microbial and enzyme activity, clay minerals and water-logging. Again, the predominant iodine forms in soils are iodide and iodate and they will partition between the solid, liquid and gaseous aggregation phases (JOHANSON 2000; JOHNSON 2003).

Important processes affecting the iodine concentration in the soil solution are the iodine **adsorption, immobilization**, **desorption**, **dissolution** and **volatilization** as well as the **leaching** from soils and the **uptake by plants** (Figure 2.4 bottom left). Studies on radioactive iodine isotopes have shown that around 70 - 90 % of the total depositions were rapidly and strongly bound in solid phases. The **adsorption** and **immobilization** of iodine on organic matter and on iron or aluminum oxides thus provides a major sink of iodine in soils (HOU ET AL. 2003; ASHWORTH 2009). The importance of humic macromolecules for binding iodine by covalent bonding was further corroborated by studies conducted by MURAMATSU ET AL. (1996) and TANAKA ET AL. (2012). In addition, filamentous fungi were found to accumulate and immobilize ^{125}I (BAN-NAI ET AL. 2006). Iodine adsorbed in solids may be transferred, by means of **desorp-**

tion and **dissolution**, to the liquid aggregation state which makes up only a small share of the total iodine in soils, with an amount of approximately 10 % (values in the range of 0.1 - 25 % were reported; JOHNSON 1980; HOU ET AL. 2003; FUGE 2005).

The **volatilization** of iodine is predominantly driven by microbial processes. Especially methyl iodide (about 40 % of the total gaseous losses) is produced by ubiquitous iodine-volatilizing bacteria, but other subordinated volatile iodine forms such as CH_2I_2, CH_2ClI, HI, HIO were also reported (AMACHI 2008; ASHWORTH 2009). Several studies have determined a relatively large range (0.004 % - 60 %) of gaseous iodine losses from soils. These wide differences were probably due to the different investigation time-spans and the different measuring methods. Indirect methods for instance, per system mass balance, noted losses in the upper range of 21 - 60 %. Direct methods, e.g. by sampling using charcoal traps and subsequent detection by neutron activation analysis (NAA), recorded gaseous losses in the lower range of 0.004 - 0.07 % (SHEPPARD AND THIBAULT 1991; SHEPPARD ET AL. 1995; BOSTOCK ET AL. 2003; ASHWORTH 2009). However, more recent studies conducted by WENG ET AL. (2009) presented volatilization losses in the range of approximately 2 - 5 %. In addition to the gaseous soil losses, plants can also convert iodine into a volatile form (CH_3I) that is emitted via stomatal pores: Depending on the investigated crop species (rice and oat) and soil conditions (oxic or anoxic) increased iodine emissions, up to 10 % of the total volatilization losses, were described in the presence of plants (MURAMATSU ET AL. 1989; MURAMATSU AND YOSHIDA 1995; LANDINI ET AL. 2012). Therefore, the total gaseous iodine emissions may vary considerably depending on the soil properties and vegetation cover, but values in the amount of < 10 % seem to be realistic. Once in the atmosphere, the different volatile iodine forms originated on land are subject to the same processes as described above for the volatile iodine forms originated from the sea. However, the volatilization of iodine from soils is subject to a large variability dependent on the organic matter content, soil texture and pH-level: Organic-rich soils, for instance, rapidly fixate iodine in the organic matter and show only low volatilization; sandy soils have a poor IFP and display correspondingly high volatilization rates. Acidic soils will be prone to a reduction of IO_3^- to I^- and subsequently to I_2, which sublimes slightly as temperature and vapor pressure increase (> 25 °C; 1 atm). Alkaline soils, in turn, will tend to show low volatilization losses by stabilizing the attendant iodate and oxidizing I^- to IO_3^- (JOHNSON 2003; FUGE 2005; WHITTEN ET AL. 2007).

The **leaching** of iodine from soils and its return to the seas and oceans though the rivers and groundwater is influenced by the IFP of soils and consequently by the content of organic matter and iron or aluminum oxides (ASHWORTH 2009). Laboratory experiments with leaching columns demonstrated iodine leaching losses in the range of approximately 3 - 4 % in a paddy soil and 5 - 6 % in a sandy soil depending on the total water quantity used to leach the columns (WENG ET AL. 2009). Moreover, the iodine form, soil pH, temperature and E_h directly or indirectly influence the leaching process. For example, iodide seems to be more mobile than iodate in soils and the predominant iodine form is dependent on the redox potential of soils: The lower the redox potential (E_h) the higher the proportion of I^- in the soil solution compared to IO_3^- that may become chemically reduced to iodide. Concordantly, soils under oxic conditions were reported to be dominated by IO_3^- in the range of approximately 75 - 90 % and soils under anoxic conditions were dominated to the extent of 60 - 90 % by I^-. Since the soil pH is directly connected to the redox potential, shifts in the pH level will influence the leaching of iodine from soils (YUITA 1992; YUITA ET AL. 2005; YAMAGUCHI ET AL. 2006). However, recent iodine leaching experiments at different pH-levels showed a dissolved iodine fraction of approximately 10 % between pH > 3 and < 9. Only a strong alkaline leaching at pH above 10 resulted in a proportion of over 30 % iodine dissolved in the soil solution (TANAKA ET AL. 2012).

The iodine **uptake by plants** has been investigated in many studies carried out with iodine isotopes (MURAMATSU ET AL. 1983, 1989; LANDINI ET AL. 2012). Although not being an essential mineral for plants, iodine in soil solutions can indeed be absorbed and embedded in vegetal tissue. Generally, the iodine uptake rates by plants appear to be low, amounting to usually less than 10 % of the total applied iodine. If supplied in low quantities up to 10 mg I $(kg\ soil)^{-1}$, iodine seems to be favorable to plant growth. Tolerance to high concentrations is situated above 50 mg I $(kg\ soil)^{-1}$ depending on the plant species (HONG ET AL. 2008; WENG ET AL. 2008a, 2008b, 2009). Although not all biochemical processes concerning the iodine uptake by plants are fully understood, some valuable suggestions have been made to date: The iodine flux across root cells may occur through putative H^+/halide transporters and anion channels (LANDINI ET AL. 2012). According to the model recently proposed by KATO ET AL. (2013), iodine applied in excess will stimulate the production of the iodate reductase enzyme in roots that subsequently converts IO_3^- to I^-. In a further step, the iodide transporter, situated in the plasma membrane of root cells, will regulate the iodide uptake in the inner cell compartments and probably the roots to shoots translocation of I^-. The same author also proposed the existence of an iodide oxidase enzyme ($I^- \rightarrow IO_3^-$) and an iodate transporter.

Once the uptake by plants is realized, rather small amounts of iodine will reach the ultimate consumers either directly by including vegetable food in the daily diet or by an intermediate step through animals and animal products. The digested and excreted iodine will then be introduced into a sewage treatment plant and thus return to the aquatic ecosystem. Finally, it should be mentioned that considerable amounts of anthropogenically produced, radioactive iodine deriving from nuclear fission fallouts, reprocessing plants and high-level radioactive waste (from leaking subterranean burial sites) have been introduced into the global iodine cycle. The leaves of food crops have been recognized in several studies as an especially relevant uptake pathway for ^{131}I and other radionuclides. Hence, radioactive iodine may ultimately accumulate in the human body and provoke serious diseases. In particular, people living in contaminated and/or iodine deficient areas are at risk of absorbing radioiodine and the importance of a well-supplied thyroid gland with stable iodine thus becomes even more evident (SINGHAL ET AL. 1998; BMU 2004; FUGE 2005; AMACHI 2008).

2.4 Iodine biofortification

The term "biofortification" describes process technologies for increasing the natural content of organic or mineral macro- and micronutrients in crops during their cultivation. Hence, the functionality of staple grains like rice, maize or wheat and several different vegetables crops is thereby improved. In addition to the research into lipids, numerous amino acids and several vitamins (e.g. vitamin A, riboflavin, ascorbic acid, vitamin E and folic acid), macro and micro elements (Ca, Mg, Fe, Zn, Cu, I and Se) are of particular research interest because they are often lacking in the human diet. Two different approaches are available to increase the content of one or more of the above mentioned nutrients in edible plant parts: **the genetic biofortification** and **the agronomic biofortification** of crops (WELCH AND GRAHAM 2004; SHAHRIARI ET AL. 2013).

The **genetic biofortification** deals, in the first instance, with conventional breeding and thus the selection of particularly efficient crops strains. Transgenic approaches are used when the natural variability of a crop for a specific trait is insufficient. LANDINI ET AL. (2012) enhanced the iodine uptake of *Arabidospis thaliana* by metabolic engineering in two different ways: Firstly, the human sodium-iodide symporter (hNIS) protein was overexpressed in different breeding lines and the more iodine-efficient (NIS16) was then selected. Secondly, the volatilization of iodine from the plant tissue was reduced by knocking-out the halide methyltransferase HOL-1. This enzyme catalyzes the methylation of iodide to methyl iodide (among other halides). Hence, the iodine uptake and its retention in crops may be improved to a significant degree thereby being advantageous to humans.

However, these techniques still require considerable basic research and to date no transgenic vegetable crop with improved traits for iodine is available. Furthermore, the acceptance of genetically modified organisms (GMO) in Europe and especially in the German population is rather low. In addition, the impoverishment of iodine in the soils of many regions of the world will constrain the effectiveness of the genetic crop improvement approach if this element is not additionally fertilized (DAI ET AL. 2004a; COSTA-FONT ET AL. 2008; HIRSCHI 2009; WHITE AND BROADLEY 2009; MORGAN 2013).

The **agronomic biofortification** of crops, hereafter referred to as biofortification, deals with the application of fertilizers and the improvement of cultivation techniques in order to increase the phytoavailability and translocation of mineral nutrients in edible plant parts (WELCH AND GRAHAM 2004; WHITE AND BROADLEY 2005; SHAHRIARI ET AL. 2013). As stated in section 2.3, plants are capable of absorbing iodine via both the root pathway and through the above ground plant parts. After its uptake by the root cells, iodine is transported mainly via the xylem and the retranslocation via the phloem is rather low (HERRET ET AL. 1962; BLASCO ET AL. 2008; VOOGT ET AL. 2010). Hence, iodine accumulation is usually higher in plant leaves than in fruits and seeds. Conditions that contribute to a high transpiration state of the leaves, e.g. the opening of the stomata under the influence of high insolation, may increase the iodine uptake as well. The iodine translocation factor (TF) has been defined as the iodine content of the vegetation divided by the iodine concentration in soils. Thus, high TF-ratios, as in the case of spinach and Chinese cabbage (TF ≥ 2.0), indicate a good transfer of iodine into the plant. Experiments on winter wheat, spring wheat and winter rye have shown a far lower TF-ratio in the range of 0.0005 - 0.02 and iodine biofortification programs were unsuccessful in most grain crops for that reason. The iodine biofortification of e.g. butterhead lettuce, spinach, Chinese cabbage, celery, carrots and radish was indeed more effective (SHEPPARD AND EVENDEN 1992; MACKOWIAK UND GROSSL 1999; SHINONAGA ET AL. 2001; DAI ET AL. 2006; TSUKADA ET AL. 2008; WENG ET AL. 2014). Hence, vegetable crops, especially leafy vegetables, might be particularly suitable for iodine biofortification programs.

The current area under cultivation with horticultural field vegetables in Germany amounts to 112.229 ha. A significant share of the total crops cultivated comprises the vegetable groups of the cabbages (e. g. kohlrabi, broccoli and white cabbage), root and tuber vegetables (e. g. carrots, onions and radish) as well as leafy (e. g. butterhead lettuce, crisp lettuce and multi-leaf lettuce) and stem vegetables (Figure 2.5; STATISTISCHES BUNDESAMT 2013, BMEL 2014). Therefore, in order to reach the greatest proportion of the German population, the biofortification of vegetables should be focused on crop species within the aforementioned groups.

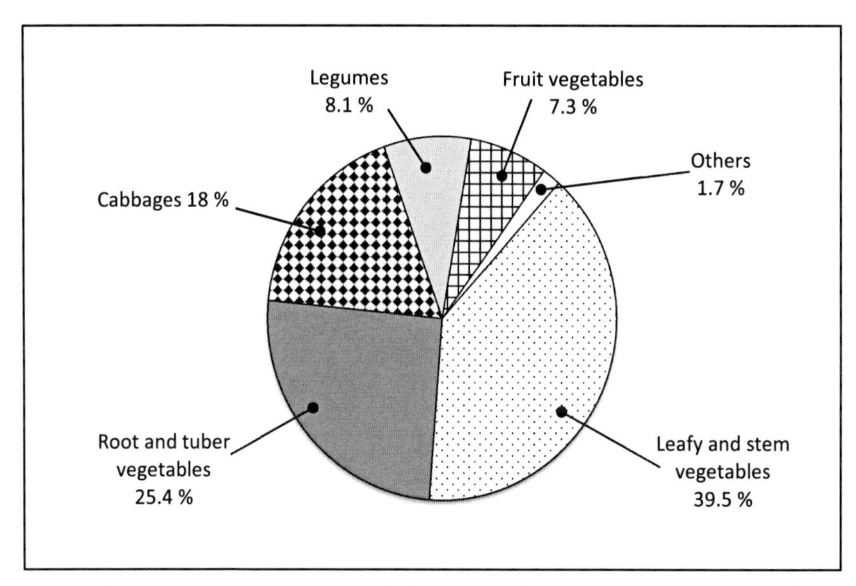

Figure 2.5 Area under cultivation with field vegetables in Germany grouped into different segments (STATISTISCHES BUNDESAMT 2013)

During the last decade, the iodine biofortification of vegetables was successfully promoted in Central Asia where vast regions are afflicted by IDD. Several studies were conducted on the soil application of iodine salts or organic iodine fertilizers e.g. kelp. Generally, the soil application of iodine was recognized to increase the iodine content of plants in concentrations proportional to the increasing fertilizer doses. Hence, this technique is considered to be a cheap and fairly reliable method for raising the iodine supply of the population by ingesting biofortified vegetables. Nonetheless, the sustainability of iodine soil applications may be rather short term thus constraining its efficiency (DAI ET AL. 2004b; HONG ET AL. 2009; WENG ET AL. 2008a, 2014).

Apart from the intensity of the applied dose, the iodine form also affects the iodine accumulation behavior of plants. Experiments with butterhead lettuce and spinach grown in nutrient solution have shown that roots absorb iodide (I^-) at a higher rate than iodate (IO_3^-). In contrast, soil applications of iodate were found to be more effective than those with iodide (ZHU ET AL. 2003; DAI ET AL. 2006; WENG ET AL. 2008a, 2008b; BLASCO ET AL. 2008), most probably because IO_3^- remains more phytoavailable in arable soils as discussed in section 2.3.

The biofortification of iodine by means of foliar applications has been poorly investigated so far. A commercial segment in the livestock production has been discovered, where iodine containing fertilizers are applied to pastures. The deposition on the vegetation of the pastures supplies the grazing cattle directly with iodine and other trace elements (RAVENSDOWN 2013, YARA 2014b). Furthermore, the reduced application quantities usually needed for foliar sprays may create an additional economic advantage compared to the soil application method (MARSCHNER 2012). However, repeated foliar spray applications are commonly needed to achieve satisfactory accumulation results. Hence, only combinations of more than one compound in each operational process, e.g. fertilizers combined with pesticides, are generally economically acceptable. In addition, only little is known about the miscibility of iodine salts with commercial agrochemicals at this point. Moreover, the low phloem mobility of iodine may be a hindering factor for the iodine accumulation in root, tuber and bulb vegetables. However, a few studies indicate that iodine is, at least in some vegetable crops, fairly phloem mobile and the foliar application approach may be a promising biofortification technique (STRZETELSKY ET AL. 2010; SMOLEŃ ET AL. 2011a, 2011b).

3 Material and methods

In the following, the common experimental procedures used in thesis, such as the chosen vegetable species, the application technique for iodine salts, the sampling and sample preparation, the standard analytical methods as well as the analytical procedures for iodine determination and the chosen statistical approaches are substantially described. Specific aspects of the individual experiments are reported in the chapters 4 and 5.

3.1 Vegetable species

The vegetable species for field trials were chosen according to two main criteria: First, the selected vegetables should be economically significant crops in Germany (cf. Figure 2.5). Second, the constitutional characteristics of the crops investigated must allow for observations on morphological/genotypic differences in the iodine accumulation behavior. Therefore, representative crop species, hereafter referred to as model crops, were selected within the group of the leafy vegetables and cabbages as well as root and tuber vegetables (Table 3.1). In addition, basil, parsley, oregano and chives were chosen as model crops for greenhouse trials, being important species in Lower Saxony within the segment of culinary herbs (NMELVL 2010).

Table 3.1 Vegetable species selected for field trials

Vegetable group	**Vegetable** (vernacular name)	**Vegetable group**	**Vegetable** (vernacular name)
Leafy vegetables	Butter head lettuce	**Cabbages**	White cabbage
	Multi-leaf butter head green		Kohlrabi
	Multi-leaf butter head red		Broccoli
	Crisp lettuce	**Root and tuber vegetables**	Carrot
	Winter spinach		Radish
	Wild rocket		Onion

3.2 Application technique for iodine salts

3.2.1 Soil application

The soil application of iodine occurred as drenches (Figure 3.1 A) at a rate of 2 L H_2O m^{-2} just before planting the peat cube transplants or sowing the seeds. The stock solutions were prepared in the laboratory with pure potassium iodide and potassium iodate salts (Ph. Eur. and Rectapur® quality, respectively; both from VWR International GmbH, Bruchsal, Germany). For this purpose, 611.34 g of KI and 801 g of KIO_3 were weighed on a precision scale (model Universal® U 4600 P, Sartorius AG, Göttingen, Germany), then transferred into 10-liter class A volumetric flasks and stirred with deionized water (0.05 µS cm^{-1}) by means of a magnetic stirrer under heat influence until complete dissolution. Once cooled, the volumetric flasks were filled with deionized water up to the mark and the stock solutions transferred into 10-liter PE-containers. To prepare the working solutions on site (Figure 3.1 B top left) an adjustable bottle-top dispenser (0 – 100 mL) was mounted on the PE-containers to allow a precise dosage of all the iodine fertilization levels, i.e. amounts of 0, 1.0, 2.5, 7.5 or 15 kg I ha^{-1}, respectively. To avoid cross contamination, separate watering cans were used for the two iodine forms and the working solutions were applied in ascending order from the unfortified control to the highest iodine dose. In order to evenly distribute the working solutions, the application was carried out in alternating directions as shown by the white and red arrows in Figure 3.1 B.

Figure 3.1 A = example of iodine application by means of soil drenches. B = application pattern of the soil drenches (white and red arrows; the black arrow indicates the watering cans and a stock solution container with dispenser)

3.2.2 Foliar application

The foliar sprays were applied once or twice at different times before the harvest of each crop. The application times differed widely (1 - 33 days pre-harvest interval) depending on the experimental issue and the vegetable species used. The working solutions were freshly prepared each week in the laboratory using the same chemicals and equipment as described in section 3.2.1. Iodine concentrations in the working solutions varied depending on the water application rate (300 - 4000 L H_2O ha^{-1}) and the chosen iodine dose (0.25 - 1 kg I ha^{-1}).

To improve the wetting of the foliage all the solutions contained 0.02 % (v/v) of the nonionic organosilicone spray-adjuvant Break-Thru® S 240 (Alzchem AG, Trostberg, Germany). In some cases, other agrochemical compounds, such as calcium fertilizers or pesticides, were mixed with iodine to test the compatibility of the different compounds.

To allow for an accurate and reproducible dosage, the iodine working solutions were portioned in PE-bottles by an automatic dispenser (model Fortuna® Optimat® 2, Poulten & Graf GmbH, Wertheim, Germany; Figure 3.2 A) in volumes and concentrations corresponding to the calculated iodine quantities per plot. The single plot portions were then sprayed according to the pattern shown in Figure 3.2 B using a hand-held spray system (model Easy-Sprayer Plus®, Lehnartz GmbH, Remscheid, Germany). This commercial airless spray gun works on the principle of oscillation caused by the alternating current driven through an electromagnetic device. The oscillating armature transmits motion to a pump piston generating suction and pressure strokes, which first draw solution from a storage container and then expel the solution at a rate of 170 mL min^{-1} through the spray nozzle producing a fine mist (LEHNARTZ 2011). For a feasible field application, the spray gun was supplied with a battery (12 V; 26 A) and an inverter system (12 to 220 volt; both from Reichelt GmbH, Sande, Germany).

During the foliar application of potentially hazardous compounds, such as pesticides, full protective clothing consisting of rubber gloves and boots, a gas mask with a filter, a safety suit and goggles was worn (Figure 3.2 C; all components from PM Atemschutz GmbH, Mönchengladbach, Germany).

Figure 3.2 A = portioning of iodine solutions in 1000 mL PE-bottles by means of an automatic dispenser. B = foliar spray application pattern (white and red arrows) on a butterhead lettuce plot. C = field application of iodine containing pest control solutions by means of a hand held spray system supplied with a battery and an inverter

3.3 Sampling and sample preparation

3.3.1 Plant material

The sampling of plant material for each plant species and trial took place at a specific point in time: For example, the earliest stage for harvesting butterhead lettuce was when it had attained an average fresh head weight of 350 g. If the requirements of the current market situation change, for instance when food retailers are short of stock, product weights less than the ideal fresh head weight and/or other poorer quality standards might be tolerated by distributors. On the other hand, a vegetable grower will try to delay the harvest until the fresh head weight has attained > 600 g if the retail market is overstocked, i.e. the price for the product and the demand are very low. However, a vegetable grower will always try to sell his products as soon as possible to clear the arable land and to commence with the next crop. Hence, the average fresh head weights in butterhead lettuce trials under practical conditions may differ from trial to trial since the variables climate, operational availability and market demand affect the cultivation. Nevertheless, the unfortified control included in every trial provides a benchmark at 100 % yield. Therefore, the specific yields of the trials in this thesis are displayed by showing the relative yields in percent.

All trial plots had peripheral rows and inner rows constituting the gross plot area (Figure 3.3 A and B). The peripheral rows where used to avoid boundary effects and were not harvested. The sampling took place only in the inner rows, i.e. the net plot area. In horticultural investigations, a widely used sampling approach is the irregular sampling with non-systematic patterns such as the “N”, “S”, “W”, “X” and other patterns (PAETZ AND WILKE 2005). In this thesis a “W” or a zigzag sampling pattern, as shown in Figure 3.3 A and B, was used for butterhead lettuce, kohlrabi, white cabbage, broccoli and crisp lettuce. All other species were harvested on a 1 m² basis of the net plot area (Figure 3.3 C, D, E and F).

The cooler early morning hours, from 5:00 - 8:00 AM, were preferred as the time of day for harvesting. The plant species harvested on trial plots using the “W” or zigzag sampling pattern were collected and stored separately in labelled plastic boxes. All other species were collected and stored in labelled polyethylene bags. The harvested crops were protected with packaging film as needed and cooled overnight at 4 °C in a cold storage house.

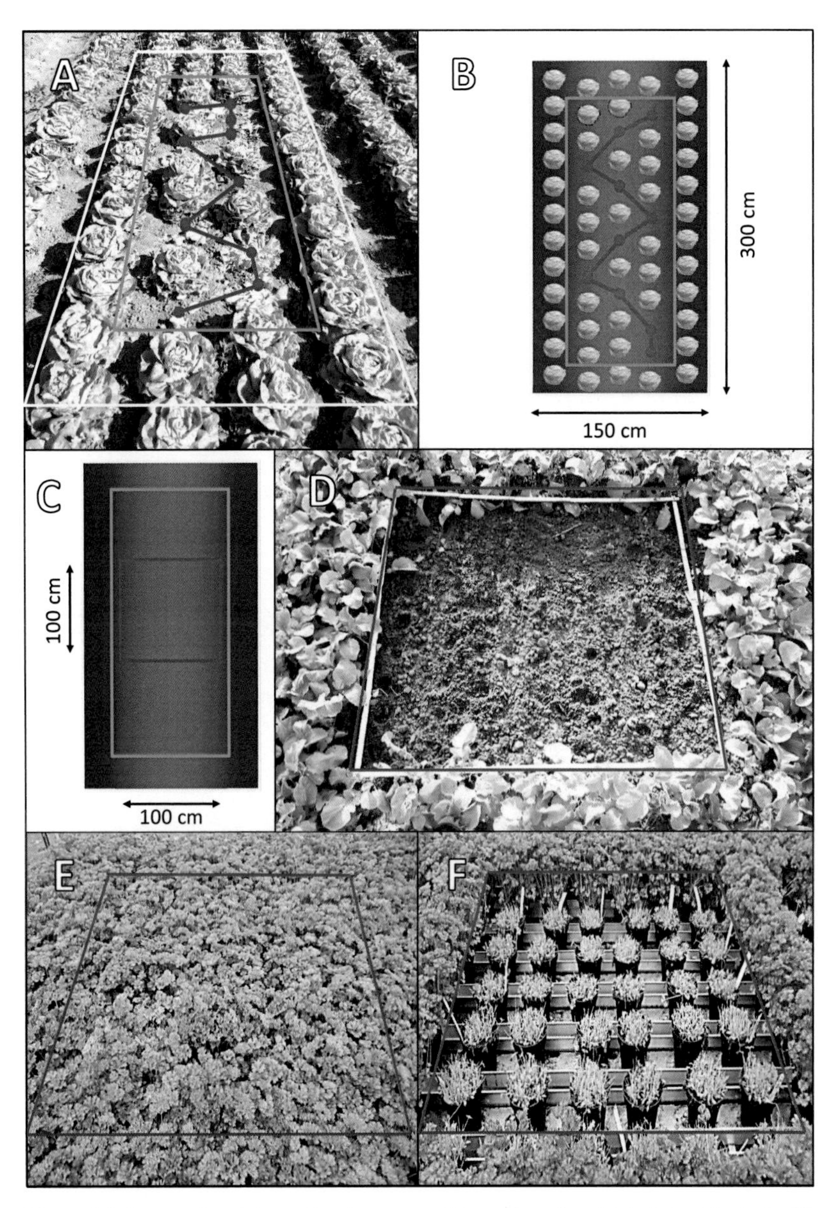

Figure 3.3 A = on site picture of a zigzag sampling pattern in a trial plot cultivated with butterhead lettuce (4 rows). B = schematic representation of the "W" sampling pattern (5 rows). C = schematic representation of sampling on 1 m² basis. D = radish sampling on 1 m² basis. E and F = parsley cultivated in greenhouse before [E] and after [F] sampling. ●—● = sampling pattern ☐ = sampling area = gross plot area ☐ = net plot area

The preparation of the plant parts occurred within 1 - 2 days after the harvest and the samples were always composite samples, i.e. consisting of crop parts from several plants per plot. The number of the harvested individuals per plot varied between 10 and several hundred depending on the vegetable species and the sampling method. The crops harvested on a one-square-meter basis, e.g. spinach, rocket, radish or the culinary herbs, had the most individuals and a composite sample was randomly extracted from the polyethylene bags to constitute a sample size of approximately 500 - 1000 g of fresh weight. Butterhead lettuce or multi-leaf lettuce were quartered by cutting the head along intersections through the center (Figure 3.4 A). Ten quarters per plot were then put together into a paper bag before drying. In the case of kohlrabi, broccoli and radish, chopped plant parts were placed onto aluminum trays for drying to avoid rot and decay. In order to conduct a differentiated plant part analysis, the respective plant parts were separated as needed (Figure 3.4 B, C, D and E).

The composite samples were washed thoroughly with tap water and the excess water was removed using a salad spinner. When investigating possible iodine salt residues on the leaf surfaces, an aliquot of one sample was separated and not subjected to the rinsing treatment.

After the preliminary preparations, the samples were transferred into a desiccating oven with air recirculation and dried at 60 °C until weight constancy for approximately 72 - 96 hours depending on the total fresh mass and the processed plant species. The fresh and dry masses were recorded before and after the drying process, to determine the dry matter fraction in percent.

The dried plant material was finely ground by using a 500 µm sieve in an ultra-centrifugal rotor mill (model ZM 100, Retsch GmbH, Haan, Germany). In some cases, bigger plant pieces, as in the case of kohlrabi or broccoli, were pre-chopped to allow easier insertion into the mill funnel. To avoid cross contamination, the samples were milled in ascending order from the unfortified control to the highest iodine dose and, before switching to the next treatment, the sieve, the recipient and the lid of the mill were rinsed thoroughly with deionized water, wiped with acetone and again rinsed with deionized water. The finely ground plant material was stored in 250 mL plastic boxes at room temperature in a storage room with no direct sunlight. Just before chemical digestion, the samples were dried again overnight at 60 °C in a desiccating cabinet and re-cooled to room temperature.

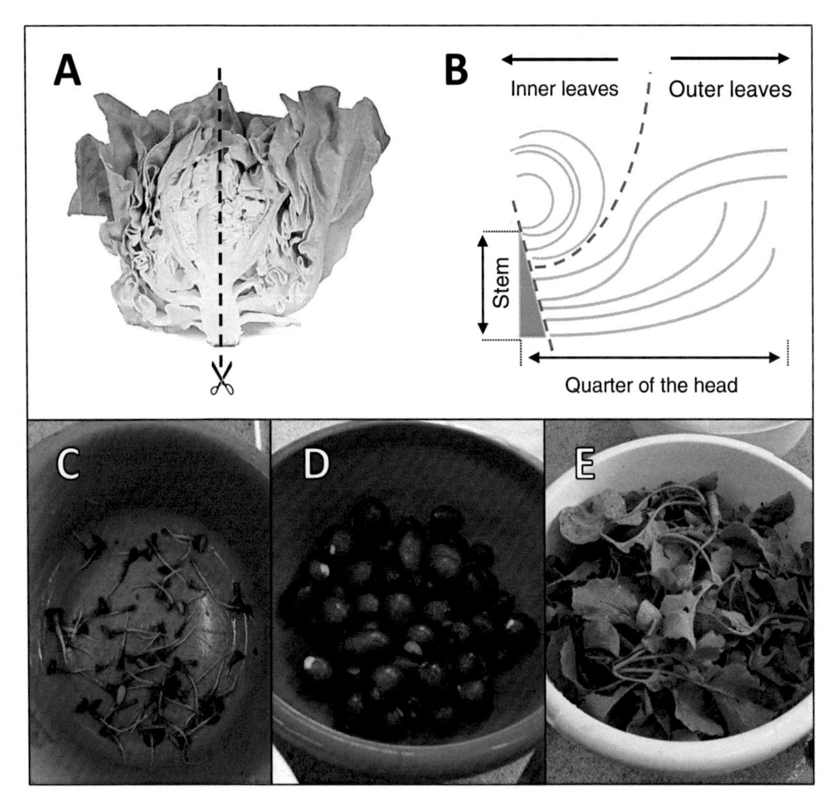

Figure 3.4 Examples for the sample preparation. A = butterhead lettuce division into quarters for a composite sample. B = schematic side view of butterhead lettuce partitioning into inner leaves, outer leaves and stem. Division of radish plants in root [C], hypocotyl bulb [D] and foliage [E]

3.3.2 Soil samples

Soil sampling in the iodine soil fertilization trials took place just before crop planting and at several intervals after the cultivation period. The sampling pattern approach was analogous to the "W" or zigzag plant sampling patterns described in section 3.3.1. The soil core samples were taken with a gouge auger model "Pürkhauer" up to a maximum depth of 90 cm in 30 cm increments. Approximately 500 - 800 grams of disturbed soil were collected per plot and depth increment. A certain degree of automation on non-stony fields was achieved by using a demolition hammer (model GBH 7-46 DE, Robert Bosch GmbH, Stuttgart, Germany) powered by a current generator available on site. For this purpose a Bosch SDS-MAX® drill bit was adapted to a connection bit accommodating the gouge auger (Figure 3.5 A). The hammering mode was used to drive the auger into the soil (Figure 3.5 B) and the rotary mode was used to carefully extract the samples from the ground. The collected samples were stored in polyethylene bags, transported in a cool box and then frozen at -18 °C. Before chemical analyses took place, the soil samples were thawed at room temperature and sieved through a 2 mm screen.

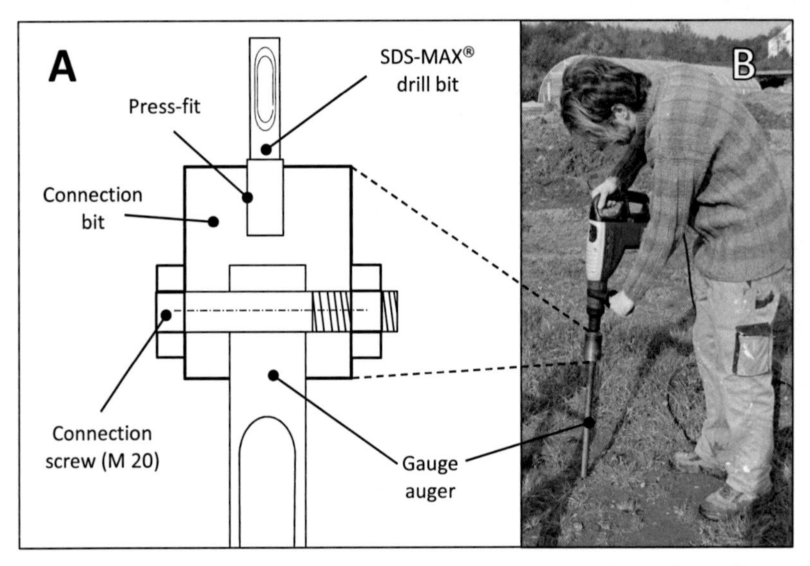

Figure 3.5 A = schematic diagram of the connection bit accommodating the Pürkhauer gauge auger (not to scale). B = automated on site core sampling using a demolition hammer (hammering mode)

3.4 Standard analytical methods

For the preparation of the analytical solutions, class A glassware was used. All the chemicals were purchased, unless otherwise stated in the text, from Merck KGaA, Darmstadt, Germany. The equipment used and the reference methods are indicated in the respective sections.

3.4.1 Soil pH

The soil pH was determined following the DIN ISO 10390 method (DIN 2005a). 20.00 g of air-dried soil (48 hours at 20 °C) were suspended in 50 mL of a 0.01 *M* $CaCl_2$-solution and stirred twice within one hour. Standard pH buffers were used for the calibration (pH 7.00 and 4.01) of a pH-meter with a combined glass electrode (Profiline® i1970 IP 66 and SenTix® 42, WTW GmbH, Weilheim, Germany). The duplicates of the soil samples were measured under constant stirring, providing a stable reading.

3.4.2 Organic matter

The soil organic matter content was determined by loss on ignition according to the DIN 19684-3 method (DIN 2000). 5.00 g of homogenized and oven-dried soil samples (dried at 105 °C until weight constancy) were weighed into porcelain crucibles and heated for 16 h at 550 °C in a muffle furnace (model Heraeus M 104, Heraeus Holding GmbH, Hanau, Germany). After cooling to room temperature in a desiccator, the crucibles duplicates were re-weighed and the organic matter content calculated in percent.

3.4.3 Determination of soluble N, P and K in soil matrix

The nitrate content of the soil was extracted at a ratio of 1 + 4 (m + v) according to the N_{min} method (VDLUFA 1997). 150.00 g of soil, at actual field moisture levels, were suspended in 600 mL of a 0.0125 *M* $CaCl_2$-solution in 1000 mL PE-bottles and stirred mechanically for one hour on a reciprocal motion shaker (model Laboshake® LS 500, C. Gerhardt GmbH & CO. KG, Königswinter, Germany). The soil sample duplicates were then filtrated through a folded filter (type MN 619 G ¼, Macherey-Nagel GmbH & Co. KG, Dueren, Germany) and the first 100 mL of the percolate were discarded.

The soluble P and K in the soil matrix were extracted according to the CAL extraction method (VDLUFA 1997). 10.00 g of air-dried soil (48 hours at 20 °C) were suspended in 200 mL of a CAL extraction solution; the CAL-solution contained 0.05 *M* calcium lactate, 0.05 *M* calcium acetate and 0.3 *M* acetic acid per liter and was buffered at pH 4.1. The soil sample duplicates were stirred mechanically in 500 mL PE-bottles for 90 minutes on a reciprocal motion shaker, filtrated through a folded filter (vide supra) and the first 20 mL of the percolate were discarded.

The nitrate and the phosphorous detection were performed using an ion chromatography system model 850 Professional® with a Metrosep® A 5 100/4.0 anion column, a conductivity detector, an automated sampler and the MagIC Net® software (all components from Metrohm AG, Zofingen, Switzerland). K was detected by means of an atomic emission spectrometer (model Elex® 6361, Eppendorf AG, Hamburg, Germany) running on the propane gas mode (1900 °C). Both detection systems were calibrated diluting the standard stock solutions of the respective element in the used extraction matrix.

3.5 Analytical procedures for iodine determination

3.5.1 Digestion of plant matrix

Iodine determination requires the decomposition of the matrix and is particularly delicate because of the high volatility of iodine. Thus, many different methods, including distillation, combustion, dry, wet, acid, alkaline, low or high temperature procedures, have been proposed so far for the digestion of biological matrices designated for iodine detection (SCHÖNINGER 1955; MERZ AND PFAB 1969; JOPKE ET AL. 1997; KNAPP ET AL. 1998; KUČERA AND KRAUSOVA 2007; KUČERA 2009). If the Sandell-Kolthoff reaction is utilized for iodine detection (see section 3.5.2), then the alkaline ashing procedure is most commonly used (AUMONT AND TRESSOL 1986). With regard to the chemism requirements of the detection system, a simplified alkaline ashing method was adapted from the procedures described by JOPKE ET AL. (1997) and KUČERA AND KRAUSOVA (2007) to convert all the present iodine forms into iodide. The main modifications to these methods were to lower the maximum ashing temperature to 550 °C and the use of KOH solely as an ashing aid instead of mixtures, e.g. KOH with $ZnSO_4$ or NaOH with Na_2O_2. In Table 3.2 a detailed instruction sheet for the digestion of plant material designated for iodine detection is shown.

Briefly, the plant material was weighed in Sigradur® glassy carbon crucibles (type GAT 4, HTW GmbH, Thierhaupten, Germany) and the KOH solution (Emsure® 47 % v/v) added. The crucibles were covered with a watch glass and then subjected to a stepwise heating procedure: Up to a maximum of 300 °C was achieved on a Trio-Term® precision aluminum hot plate and then up to 450 °C on a Ceran® hot plate (model C 450, C. Gerhardt GmbH & CO. KG, Königswinter, Germany). Subsequently, the crucibles were placed in a muffle furnace at 550 °C and, after cooling, the fusion cake was solubilized by adding deionized water and placing the crucibles in an ultrasonic bath (model Sonorex® RK 255 H, Bandelin electronic GmbH & Co. KG, Berlin, Germany). The resulting solution was then quantitatively transferred in volumetric flasks by rinsing the crucibles and the watch glasses with deionized water.

Table 3.2 Instruction sheet for the alkaline fusion procedure

Step	Action	Explanatory remarks and needed material
1	Weigh 0.1000 g ± 0.0005 g of dried and ground plant material in the crucibles	Precision scale; micro-spoon; paintbrush and paper towels; glassy carbon crucibles
2	Pipet 1,618 mL of the KOH solution (47 %; ρ = 1.4755 g cm^{-3}) in the crucibles	Transfer-pipette; PP-Tips; do not touch the crucibles or the plant material with PP-tips; dose cautiously!
3	Place the crucibles on the precision hot plate and cover them with a watch glass; position the watch glass slightly skewed, in a way to allow a slot formation (the reaction intensity will thereby be reduced); heat stepwise at 100, 150, 200 and 300 °C for approx. 10 minutes, respectively	Watch glasses; wear gloves and safety goggles; turn the fume hood on; a strong reaction and foaming of the plant material is possible; allow the samples to react until complete cessation
4	Remove the watch glasses with tweezers and place them upside-down in a row next to the hot plate; allow the substance to vaporize at 300 °C until a brownish melt is formed	Approx. 20 minutes; close the fume hood; tweezers; remove and replace the watch glasses carefully!
5	Place the watch glasses on the crucibles and transfer them with the crucible tongs to the hot plate, heat up to 450 °C	Crucible tongs; allow the KOH to melt for approximately 10 minutes.
6	Place the crucibles with watch glasses in the muffle furnace and heat them at 550 °C for 5 minutes	Pre-heat the muffle furnace at 580 - 590 °C; the temperature will drop whilst opening the door!
7	Take the crucibles out of the muffle furnace and allow them to cool down	For approx. 20 minutes at room temperature
8	Half-fill the crucibles with deionized water and place them into the ultrasonic bath	For approx. 1 minute
9	Transfer the solubilized fusion cake quantitatively in 100 mL volumetric flasks by rinsing the crucible and the watch glass	All surfaces which came into contact with the fusion compounds must be thoroughly rinsed!
10	Rinse the cones with deionized water and fill the volumetric flasks up to the mark	Insert the stopper and stir well
11	Filter the solution through a folded filter type MN 619 G ¼ and collect the percolate in 100 mL PE bottles	Discard the first 20 mL of the percolate; a clear solution must be obtained!

3.5.2 Iodine detection in alkaline matrix

The iodine detection in aqueous solutions after alkaline digestion was performed according to the Quick-Chem® method 10-136-09-1-A (SWITALA 2001) using a flow injection analysis (FIA) system model Quick–Chem® 8500 equipped with an automated diluter and sampler, an iodide manifold and the Omnion® 2.2.2 software (all components from Lachat Instruments, Hach Company, Loveland, CO, USA). This automated system is based on a catalytic spectrophotometric method which exploits the Sandell–Kolthoff reaction: Iodide catalyzes the discoloration of yellow ammonium cerium (IV) sulfate by reduction with arsenic (III) oxide in an acid medium (SANDELL AND KOLTHOFF 1934, 1937; As_2O_3 from Sigma-Aldrich Chemie GmbH, Schnelldorf, Germany). The iodide manifold used (Figure 3.6 A) follows the inverse absorbance at 420 nm, which is proportional to the iodide concentration in the samples; NaCl is loaded in surplus into the flow path to prevent possible interferences caused by Cl^- ions and the carrier injection (0.2 *M* KOH) provides a baseline identical to the digestion matrix (JOPKE ET AL. 1996; SWITALA 2001).

The major advantages of the FIA were the low cost of the equipment required, the low detection limit and the high sample throughput. On the other hand, two main problems arose during the course of the detection runs and the system proved to be high-maintenance: Firstly, the chromatograms showed spikes from the bottom of the inverse peaks substantially disturbing the integration (Figure 3.6 B). This occurrence was most probably due to carbonates generated by the combustion of organic C from plant material and air-CO_2 during the alkaline digestion ($2\ KOH + C_{org} + O_2$ and/or $2\ KOH + CO_2 \rightarrow K_2CO_3 + H_2O\uparrow$). The carbonaceous samples reacted in turn, after being injected in the iodide manifold, with the acidic reagents ($K_2CO_3 + H_2SO_4 \rightarrow K_2SO_4 + H_2O + CO_2\uparrow$) and carbon dioxide bubbles formed disturbing the course of the chromatogram when passing through the flow cell. Attempts to eliminate the carbonates in the samples by adding acid reagents before the FIA according to WELSHMAN ET AL. (1966) were rejected after observing similar high iodide losses as reported by FOSS ET AL. (1960). Further attempts trying to exclude carbonates by the addition of barium hydroxide and the subsequent filtration of the barium carbonate precipitates ($K_2CO_3 + Ba(OH)_2 \rightarrow 2\ KOH + BaCO_3\downarrow$) resulted in excessive blank levels and were also rejected. The Sandell–Kolthoff reaction can be affected by other ions, apart from Cl^-, such as NO_2^-,

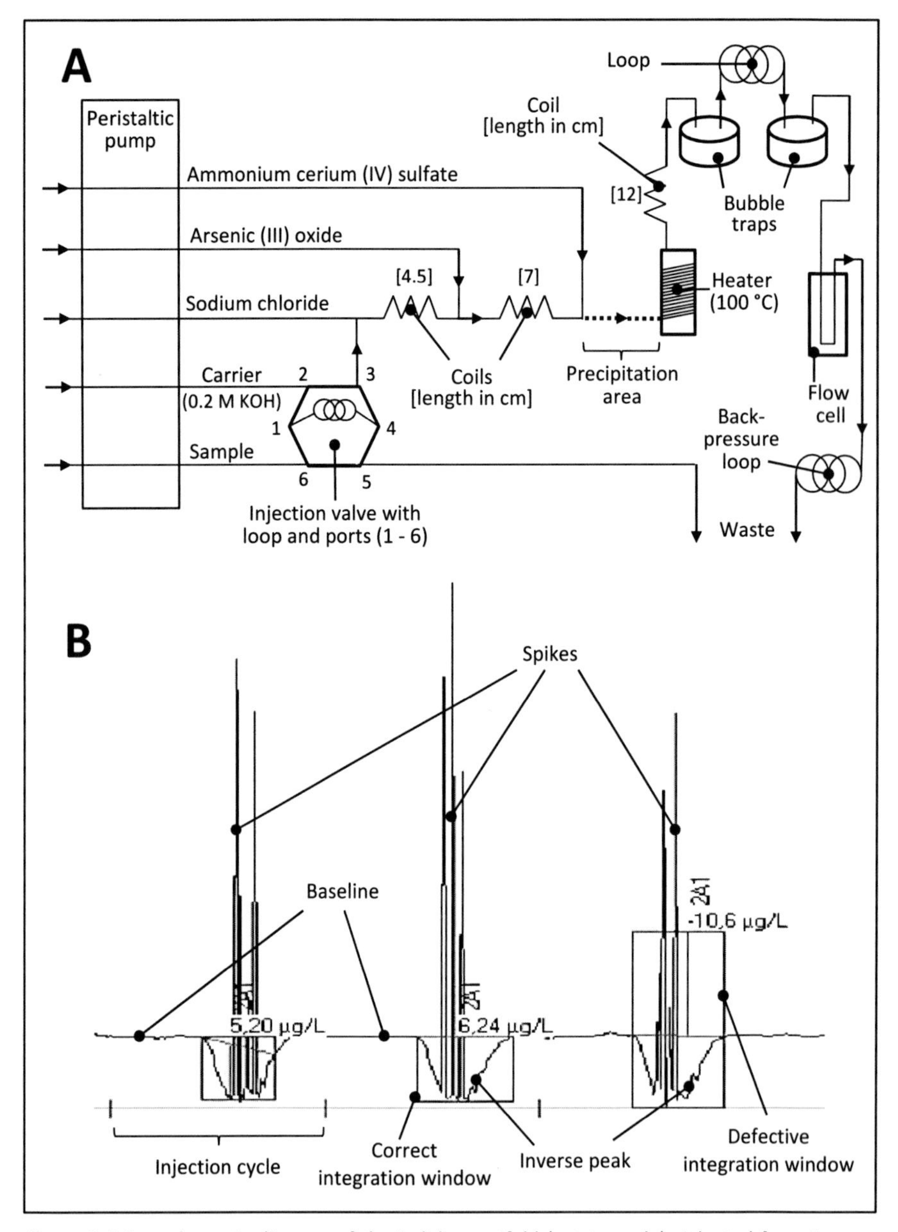

Figure 3.6 A = schematic diagram of the iodide manifold (not to scale). Adapted from SWITALA (2001). The dotted line indicates the cerium carbonate precipitation area. B = example of chromatogram interferences (spikes) caused by CO_2-bubbles

SCN^-, Fe_2^{+}, BrO_3^-, MnO_4^-, CN^- and Hg^+ (JOPKE ET AL. 1996). Hence, the sample processing with barium hydroxide, the eventually contained impurities in the final solution as well as its properties as an alkaline agent, may have led to an altered chemism and a consequent bias of the chromatogram.

However, the interferences caused by CO_2-bubbles could almost entirely be removed by the installation of two bubble traps (Omnifit® part no. 006BT; Diba Industries Ltd., Cambridge, UK; Figure 3.6 A top right).

The second problem was an increasing accumulation of precipitates located in the tubing between the last T-connection and the heater unit (dotted line in Figure 3.6 A). The precipitation was probably caused by a reaction of the cerium (IV) sulfate with the carbonates contained in the samples, e.g. $6\ CO_3^{2-} + 6\ Ce(SO_4)_2 \rightarrow 6\ SO_4^{2-} + Ce(CO_3)_6^{8-}\downarrow$ or other Ce(IV) complexes like $Ce(CO_3)_5^{6-}$ (LARABI-GRUET ET AL. 2007). The formation of cerium carbonate in the tubing, barely soluble at room temperature, probably led to a flow blockage and eventually to a detection run break-down. This occurrence could be eliminated to a high degree by injecting an acidic flush solution (10 % H_2SO_4 w/v) after every 2 - 3 sample injections (WELSHMAN ET AL. 1966; MATTHES ET AL. 1978).

Nevertheless, a proper function of the FIA system could only be achieved through frequent servicing, including the weekly replacement of peristaltic pump tubing, fitting O-rings, manifold tubing, bubble trap membranes, and, the cleaning of all fittings and the flow cell by ultrasonication in an acidic solution (vide supra).

3.5.3 Quality assessment and quality control of iodine determination in plant matrix

A number of arrangements were implemented in order to achieve analytical data of an adequate quality for the research purpose of this thesis: Firstly, the field and greenhouse trials were performed with unfortified control treatments and 3 - 5 replicates per treatment. Randomized sampling was performed as described in section 3.3.1. The preliminary sample preparation steps and the digestion procedures were conducted under guidance and by means of instruction sheets (see Table 3.2). Method blanks and reference material duplicates were added to every digestion process (Fapas® test material no. T07110; The Food and Environment Research Agency, York, UK).

The calibration standards were prepared by means of an automated dilution unit (model PDS 200, Lachat Instruments, Hach Company, Loveland, CO, USA) using a certified iodide stock solution (CRM 27031; lot N RTA 25; 1000 mg I L^{-1} ± 0.2 %; Carl Roth GmbH + CO. KG, Karlsruhe, Germany). Furthermore, the calibration curve was fitted in the range of 0 - 10 µg I L^{-1} (0, 2, 4, 6, 8 and 10 µg I L^{-1} calibration row) using a linear curve fit following the DIN 38402 method (DIN 1986; part 51). The range between 10 - 100 µg I L^{-1} (10, 20, 30, 40, 50, 60, 70, 80, 90 and 100 µg I L^{-1} calibration row) was fitted by a 2nd degree polynomial curve following the DIN ISO 8466-2 method (DIN 2004). The split calibration procedure had the main advantages of improving the accuracy and the precision in the low detection range and, at the same time, broadening the detection range between two calibration decades. Some selected calibration standards were additionally run as unknown samples (check standards) to verify the current detection run.

In order to run a method comparison, three subsets of samples, including two sets with very low and low iodine levels and one set with a broad amplitude in the iodine concentration, were sent to an accredited laboratory (Thüringer Umweltinstitut, Henterich GmbH & Co. KG, Krauthausen, Germany) to determine the iodine content according to the DIN EN 15111 method (DIN 2007a), i.e. plant matrix digestion by TMAH incubation and detection by ICP-MS. The values were then compared to the findings in this study by a correlation analysis. The results showed a very poor correlation in samples at iodine levels below 1 µg I L^{-1} (Figure 3.7 A). A slightly better correlation and recovery were observed in a concentration range up to approximately 2 µg I L^{-1} (Figure 3.7 B). However, the best conformity was found in samples with the broadest iodine concentration range (approx. 2 - 190 µg I L^{-1}) showing a good coefficient of determination (0.9983) and a slope (0.9251) close to 1 (Figure 3.7 C). This conversely means a recovery rate of 92.5 %, which is comparable to the findings of Johner et al. (2012). That implies reliable iodine detection concentrations between 2 and 100 µg I L^{-1} (higher concentrations were diluted to within the calibration range).

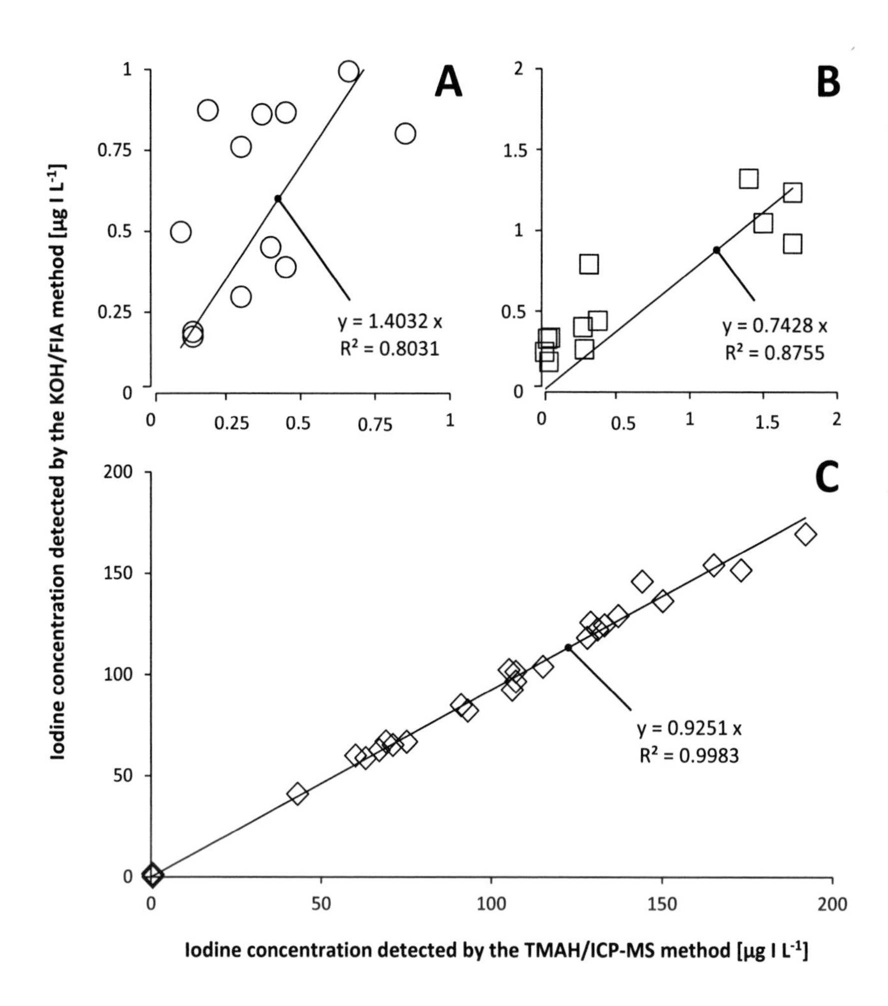

Figure 3.7 Correlation between the KOH/FIA and the TMAH/ICP-MS detection methods in sample subsets with very low [A, circles], low [B, quadrates] and very broad [C, diamonds] iodine concentration levels. (T-test [A]: probability level = 0.0000, power = 0.9557, T-value = 6.6989, n = 12; T-test [B]: probability level = 0.0000, power = 0.9992, T-value = 8.7935, n = 12; T-test [C]: probability level = 0.0000, power = 1.00, T-value = 124.2553, n = 28). $\alpha = 0.001$

Considering the desirable target levels of iodine accumulation in edible plant parts [50 - 100 µg I (100 g FM)$^{-1}$] and the mean dry matter fraction of butterhead lettuce (approximately 4 %) for instance, a minimum detection amplitude between 12.5 and 25 µg I L^{-1} is required and therefore lies comfortably within the reliable range.

On the other hand, plant samples with very low iodine content (< 2 µg I L^{-1}), like the unfortified controls, may show high bias and noise levels resulting in less significant differences than are actually present in a low concentration range. However, the interlaboratory comparison demonstrated that the iodine detection method used in this thesis is acceptable, being precise and accurate within the needed working range.

3.5.4 Extraction and detection of iodine in soil matrix

The calcium-chloride extractable iodine fraction in soil was performed according to the method suggested by ALTINOK ET AL. (2003). This method is analogous to the N_{min} method described in section 3.4.3 (VDLUFA 1997). Due to an incompatible matrix, the iodine contents in calcium chloride extracts could not be detected by FIA and were therefore determined according to the DIN EN ISO 17294-2 method (DIN 2005b) by an external laboratory (Thüringer Umweltinstitut, Henterich GmbH & Co. KG, Krauthausen, Germany).

3.6 Statistical procedures

In the agricultural field experimentation many scientists choose the approach of the randomized complete block design (RCBD) which especially suits trials with a small number of treatments and a predictable productivity gradient (GOMEZ AND GOMEZ 1984).

The clear advantages of a RCBD are the increased experimental precision due to reduced experimental error, the simple structure and the relatively easy statistical analysis even when some values are missing. However, in some cases, more complicated designs such as the split-plot design might be very useful for testing one factor more severely than another since it provides an increased precision in the estimates of the sub-plot treatments (RANGASWAMY 2010). In other cases, the adaption of the experimental design to the conditions of the site or to some of the operational events, such as the tillage, is inevitable. For some of the aforementioned reasons, the experimental models in this thesis are specifically tailored to each case. An overview of the chosen experimental models is given in Table 3.3. The first soil application trials (W1, W2 and H in season 2010 and later the W5 and W6 trials in season 2011) were conceived as a split-plot design because of two main considerations: Firstly, the trials had to be positioned in parallel, in such a manner as to allow practicable tillage operations, i.e. strip tilling with rotary tillers in the case of an analysis over years. Secondly, the parallel positioning of two or more vegetable species would have allowed for the option of a comparison between species and/or between the species over time, if results would have shown that the data needed further analysis. Contemporaneously, the required iodine fertilizer quantity would have been tested more severely than the iodine form.

In the case of trial P, the adaption of a long-term trial to test the influence of different pH-levels on vegetables, started in the 1960s years, was made possible by modifying the existing RCBD to a strip-plot design. This had two major advantages: Firstly, an easy adaption to the on-site conditions and secondly, a weighted estimation of factors and interactions, which would test the interaction (pH-level x iodine form) more severely than the factors. However, the RCBD approach best suited most of the trials in the seasons 2011 - 2012 on account of the changing experimental emphasis based on new results. Hence, the estimation of the factors and their respective interactions with the same precision was more desirable. Furthermore, the statistical computation was easier and the trial installation slightly less laborious.

Table 3.3 Overview of the chosen trial models. ANOVA models were adapted from Gomez and Gomez (1984)

	Design	Substantiation	ANOVA model
Trials W1 – W6, H	Split-plot	Increased precision in the estimates of the sub-plot treatments; adaption to the tillage operations	see ANOVA split-plot below
Trial P	Strip-plot	Adaption to existing trial and higher precision in measuring the A x B interaction effect	see ANOVA strip-plot below
All other trials	RCB	Increased experimental precision and easy computation	see ANOVA RCB below

Trials W1 – W6, H – Split-plot

Source of variation	Degree of freedom	Sum of squares	Mean square	Computed F	Tabular F
Block	$r-1$				
Main plot factor (A)	$a-1$				
Error (a)	$(r-1)(a-1)$				
Subplot factor (B)	$b-1$				
A x B	$(a-1)(b-1)$				
Error (b)	$a(r-1)(b-1)$				
Total	$rab-1$				

Block

Main plot factor

Subplot factor

14	11	13	12	10	21	22	24	23	20

Trial P – Strip-plot

Source of variation	Degree of freedom	Sum of squares	Mean square	Computed F	Tabular F
Block	$r-1$				
Horizontal factor (A)	$a-1$				
Error (a)	$(r-1)(a-1)$				
Vertical factor (B)	$b-1$				
Error (b)	$(r-1)(b-1)$				
A x B	$(a-1)(b-1)$				
Error (c)	$(r-1)(a-1)(b-1)$				
Total	$rab-1$				

Block

Horizontal factor

Vertical factor

51	21	11	41	31	50	20	10	40	30	52	22	12	42	32
52	22	12	42	32	51	21	11	41	31	50	20	10	40	30
50	20	10	40	30	52	22	12	42	32	51	21	11	41	31

All other trials – RCB

Source of variation	Degree of freedom	Sum of squares	Mean square	Computed F	Tabular F
Block	$r-1$				
Treatment	$ab-1$				
[Factor A]	$[a-1]$				
[Factor B]	$[b-1]$				
[A x B]	$[(a-1)(b-1)]$				
Error	$(r-1)(ab-1)$				
Total	$rab-1$				

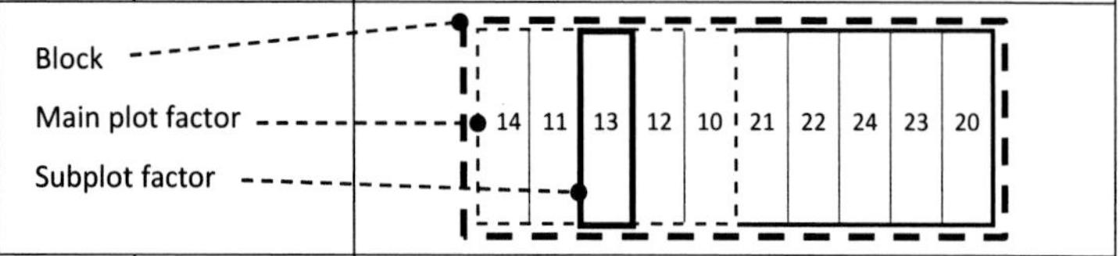

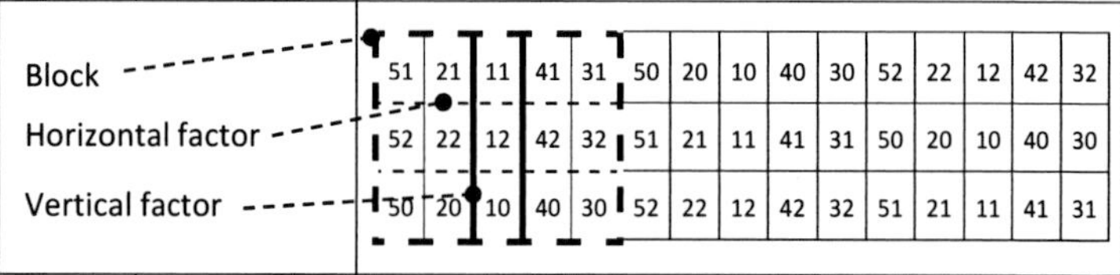

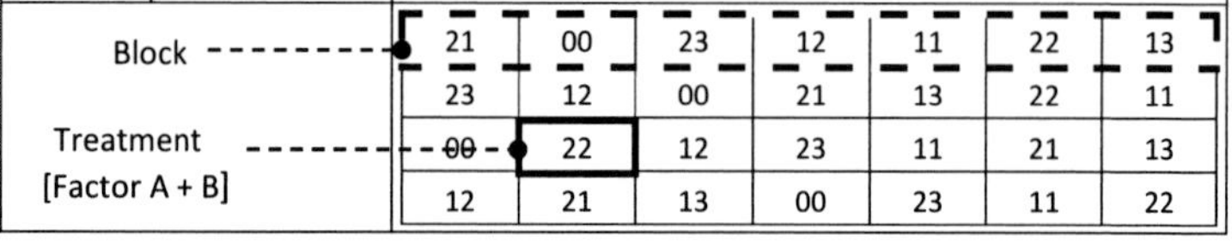

Block

Treatment [Factor A + B]

21	00	23	12	11	22	13
23	12	00	21	13	22	11
00	22	12	23	11	21	13
12	21	13	00	23	11	22

The analysis of variance (ANOVA) requires some mathematical assumptions, i.e. the additivity of the environmental and treatment effects and that the experimental errors are independent, normally distributed, and have common variance. By using proper trial randomization, created by a random number generator (Microsoft Excel® 2010), the assumption of the independence of errors can be implied and does not have to be discussed any further (LITTLE AND HILLS 1978, GOMEZ AND GOMEZ 1984, PEARCE ET AL. 1988).

The other assumptions have been tested using the Number Crunching Statistical System (NCSS® 2007): A one-way ANOVA and the incorporated preliminary normality tests were performed setting the iodine content as a response variable and the different consecutively numbered treatments as fixed factor variables. The modified Levene equal variance test and the D'Agostino Skewness, Kurtosis and Omnibus tests were conducted at the 5 % significance level ($\alpha = 0.05$) on raw data in order to detect violations on the normality assumptions of the residuals (D'AGOSTINO ET AL. 1990; HINTZE 2007). The trials conducted in this thesis had different sample sizes ranging between $n = 28$ and 45 depending on the number of the treatments (7 - 10) and their respective repetitions (3 - 5). Although sample sizes $n \geq 30$ are not likely to cause major problems even violating the normality assumptions (GHASEMI AND ZAHEDIASL 2012), parametric tests were only used if the dataset displayed a normal distribution.

To normalize the datasets which did not meet these assumptions, or to minimize the coefficient of variation (CV) of a few normal datasets with a CV higher than 25 %, the data transformation was performed either using a logarithmic or a square root transformation in one of the following manners:

$$x' = \mathrm{Log}_{10}\, x \quad \text{or} \quad x' = \mathrm{Log}_{10}\,(x + 1)$$

$$x' = \sqrt{x} \quad \text{or} \quad x' = \sqrt{(x + 0.375)}$$

The logarithmic transformation was the first choice because of the wide range of data values in most of the datasets (0.x to several hundreds) and the multiplicative effects stated in some trials (multiplicity turned to additivity). Since this transformation cannot be applied to sets with zero values, the number 1 was added to datasets containing 0.x values (LITTLE AND HILLS 1978). In cases where logarithmic transformation, either with or without a constant number, did not normalize the dataset, the variance stabilizing square root transformation was performed; being a good transformation for Poisson distributions. In some cases 0.375 as constant number was added since it fits means with small values (fewer than 5) better (OTT 1988). If normality assumptions could not be met by using the common transformations mentioned above, the more laborious Box-Cox Transformation (BOX AND COX 1964) was applied to skewed datasets using the following formula:

$$x' = \frac{x^{\lambda} - 1}{\lambda} \quad ; \quad \lambda \neq 0$$

$$x' = \ln x \quad ; \quad \lambda = 0$$

In order to detect the applicable lambda value for a specific dataset, an algorithm had to be applied for the value estimation. Due to the fact that NCSS® version 2007 does not incorporate the Box-Cox routines, SPSS® version 20 (IBM® SPSS® 2012) was used for this purpose computing the following syntax commands (modified, adapted from OSBORNE 2010):

```
COMPUTE var1=oregano-.17.
execute.
VECTOR lam(41)/ xl(41).
LOOP idx=1 TO 41.
- COMPUTE lam(idx)=-2.1 + idx * .1.
- DO IF lam(idx)=0.
- COMPUTE xl(idx)=LN(var1).
- ELSE.
- COMPUTE xl(idx)=(var1**lam(idx) - 1)/lam(idx).
- END IF.
END LOOP.
EXECUTE.
FREQUENCIES VARIABLES=var1 xl1 xl2 xl3 xl4 xl5 xl6 xl7 xl8 xl9 xl10 xl11 xl12 xl13 xl14
xl15xl16 xl17 xl18 xl19 xl20 xl21 xl22 xl23 xl24 xl25 xl26 xl27 xl28 xl29 xl30 xl31 xl32 xl33 xl34
xl35 xl36 xl37 xl38 xl39 xl40 xl41
/format=notable
/STATISTICS=MINIMUM MAXIMUM SKEWNESS
/HISTOGRAM
/ORDER=ANALYSIS.
```

The estimation of λ was computed for 41 different lambdas simultaneously, choosing the most suitable one by approximation. The first COMPUTE anchored the variable at 1.0, while the minimum value in the dataset "oregano" was 1.17. Other dataset minima had to be anchored at 1.0 as needed. The LOOP has a limited range for variables (-2.1 to + 2.0) and examination intervals of 0.1, which were edited depending on the needs of the analyzed datasets. For example, after steps of approximate adjustments beginning at -2.1, the fine-tune analysis of the oregano dataset was performed within a range of + 0.412 to + 0.416 in intervals of 0.0001, matching a skewness of 0.000 with λ at 0.4125 (OSBORNE 2010).

Once data was scrutinized for violation of the assumptions and normalized by transformation if needed, a one-way ANOVA was performed again in the manner described above. Despite all the transformation strategies, in some cases it was not possible to normalize the data by transformation. In such occurrences, the data was transformed into ranks. An overview of the raw and transformed datasets is given in Table 3.4. The datasets transformed into ranks were analyzed by means of the parameter-free Wilkoxon rank-sum test: The different treatments of one trial were compared pairwise to the unfortified control. In some cases, when datasets of related trials where compared, the pooling of datasets led to not normally distributed data. In these cases, all the affected datasets were transformed into ranks and the trials tested using the above-mentioned parameter-free methods (cf. Table 4.8). Additionally, the Friedman's rank test was conducted if balanced data was available.

In the case where a relationship between two variables had to be estimated (e. g. Figure 3.7, section 3.5.3), a regression and correlation analysis was performed and a two-sample T-test was run at $\alpha = 0.001$ if at least $n = 12$ observations were present.

Decision making with respect to the best fitting multiple comparison procedure (MCP) was more laborious since most statistical packages offer a wide variety of post-hoc tests based upon methods suggested, among others, by DUNNETT (1955), DUNN (1964), HOCHBERG AND THAMANE (1987) and HSU (1996). NCSS® 2007 provides 10 different MCPs, 2 of which, the Tukey-Kramer and the Bonferroni procedure, are recommended for the planned comparison of all possible groups (HINTZE 2007). The Tukey-Kramer test uses the experimentwise error rate and the studentized range distribution (q) in the following formula:

$$\frac{| y_i - y_j |}{\sqrt{\frac{S^2}{2}\left(\frac{1}{n_i} + \frac{1}{n_j}\right)}} \geq q_{\alpha,k,df}$$

In contrast, the Bonferroni procedure chooses the comparisonwise error rate (α) to control the desired experimentwise error rate (α_f) with k means in the following manner:

$$\alpha = \frac{\alpha_f}{k(k-1)/2} \quad \text{and} \quad \frac{| y_i - y_j |}{\sqrt{S^2\left(\frac{1}{n_i} + \frac{1}{n_j}\right)}} \geq t_{\alpha,df}$$

The Bonferroni test considers all numbers of comparisons to be made and is less likely than other MCPs to find significant differences (CABRAL 2008). Analogous observations were made comparing the Tukey-Kramer and the Bonferroni test in the evaluation of the trials in this thesis. However, in some of these datasets a similar classification between the two test methods was observed and the decision for the best applicable MCP thus fell on the more conservative Bonferroni procedure.

If significant differences were detected by the one-way ANOVA and treatments subsequently grouped by MCPs, other points of interest, namely the experimental factors and the respective interactions, were analyzed performing an ANOVA according to the general linear model (GLM) approach. Analogously to the one-way ANOVA but with more than one factor, the GLM procedure uses an additive model to explain the response variable Y, i.e. the sum of all parameters of the population. It follows the mathematical model in the next example: Y_{ijkl} = grand mean m + effect factor a_i + effect factor b_j + effect interaction $(ab)_{ij}$ + effect factor c_k + effect interaction $(ac)_{ik}$ + effect interaction $(bc)_{jk}$ + effect interaction $(abc)_{ijk}$ + error e_{ijkl} (OTT 1988, HINTZE 2007). In NCSS® 2007 the implementation of this mathematical model is slightly simplified (the grand mean m is implied and the error e_{ijk} is pooled automatically) but analogous: Generally, the block was represented by A and set as a random type factor. Interactions with A, represented by AB, AC and ABC, are of interest when split errors arise, such as main-plot errors (split-plot design) or horizontal and vertical factor errors (strip-plot design; see Table 3.3). The interaction ABC is pooled in the error (denoted with S in the output) and is not written in the following custom models. The factors B and C represent two fixed factors with different specifications, for example the iodine dose and the iodine form. The interaction of the two factors is always represented by BC. Therefore, split-plot trials were analyzed according to the custom model A + B + AB + C + BC, the strip-plot trial according to

A + B + AB + C + BC + AC and the randomized complete block trials according to A + B + C + BC.

A common unfortified control, as applied in some RCB trials of this thesis, could cause two undesired effects: Firstly, the statistical program would be prone to conceive a full factorial combination of factors, creating combinations counting no figure as a response variable and, secondly, it could falsely report significant interactions between the investigated factors due to the dependency of treatments. Therefore, some datasets of RCB trials were computed without unfortified control data in the GLM procedure.

The coefficient of variation (CV), expressing the experimental error as a percentage of the mean, is a good measure for the reliability of an experiment. Thus, the lower the value, the higher the reliability of the experiment (GOMEZ AND GOMEZ 1984). The CV of the trials was computed using the grand mean of the datasets and the mean square error (MSE) as follows:

$$\mathbf{CV\ [\%]} = \frac{\sqrt{\mathbf{MSE}}}{\mathbf{Grand\ mean}} \times 100$$

In biological data there is a great variation of the CV depending on the type of the experiment, the cultivated crop and the measured character. Although ranges between 3 - 20 % are considered as normal values, coefficients of variation above 25 % are not uncommon (GOMEZ AND GOMEZ 1984, GAUCH 1992). In Table 3.4 the experiments in this thesis are listed by code and year giving an overview of the achieved CV in dependence of the applied data transformation. In 25 out of 28 cases the CV lies within a range of 5 to 24 %. In 3 cases the CV was above 25 %, 1 of which was not altered by transformation (rank transformation occurred without altering the CV of the raw data). In addition, these outliers are based on datasets containing very small values in raw data (W3, W6 and BR V), suggesting a high bias and a high noise level in trials where approaches to iodine accumulation had no or only little effect (cf. 3.5.3). The arrangement of the coefficients of variance by year shows a slight improvement from the beginning of the experimental series to the last trials conducted in 2012. This can be attributed to an improved application technique and to the increasing number of repetitions. The overall coefficient of variation, with a grand mean of 18 %, lies within a normal range and the experimental execution of the trials in this thesis can thus be regarded as fairly reliable.

Table 3.4 Overview of data transformation and respective coefficient(s) of variation arranged by experimental year. * = including transformed datasets; ** = computation of CV not possible, CV of raw data was used instead

Trial		Raw data			Transformed data			
Code	Year	MSE	Grand mean	CV [%]	Transformation	MSE	Grand mean	CV [%]
W1	2010	112	26	41	Log_{10} (x + 1)	0.03	0.976	**20**
W2		792	76	37	Log_{10} (x + 1)	0.03	1.471	**13**
W3		21	11	41	Rank**	-	-	**41**
W4		4010	162	39	Square root (x + 0.375)	4.97	11.01	**20**
H		26	17	29	Square root	0.37	3.593	**17**
P		4275	115	57	Log_{10}	0.01	1.723	**7**
W5	2011	19	10	44	Square root (x + 0.375)	0.56	3.104	**24**
W6		5	3	63	Log_{10} (x + 1)	0.03	0.483	**40**
WS		2884	723	23	Rank**	-	-	**23**
WK		21	9	49	Log_{10} (x + 1)	0.01	0.913	**15**
EI		1118	104	32	Square root	1.67	7.463	**17**
RU		286	71	24	No transformation	-	-	**24**
KS1		2779	293	18	No transformation	-	-	**18**
KS2		248	47	33	Log_{10}	0.01	1.547	**9**
KS3		5567	317	24	No transformation	-	-	**24**
BR G		4	7	30	Square root (x + 0.375)	0.17	2.599	**16**
BR V		3	3	55	Square root (x + 0.375)	0.24	1.822	**27**
BA		1726	181	23	Square root	1.42	12.13	**10**
OR		3077	227	24	Box-Cox	0.41	5.984	**11**
PE		2041	144	31	Box-Cox	1.14	5.327	**20**
SL		1391	107	35	Square root	2.05	9.384	**15**
AT	2012	685	112	23	Log_{10}	0.00	1.791	**5**
CA		68	42	20	No transformation	-	-	**20**
PS		317	94	19	Square root	0.60	8.491	**9**
SA		494	128	17	Square root	0.81	9.938	**9**
TR		788	128	22	Square root	1.10	10.25	**10**
WA		242	81	19	No transformation	-	-	**19**
TZ		231	90	17	No transformation	-	-	**17**
CV mean arranged by year (±SD)							2010	**20*** ±11.7
							2011	**24*** ±7.9
							2012	**13*** ±5.7
CV grand mean (±SD)								**18*** **±8.6**

4 Iodine biofortification by means of soil fertilization

4.1 Abstract

The iodine biofortification of vegetables by means of soil fertilization was investigated in field experiments on sandy loam soils over two growing seasons. With a rising iodine supply (0, 1.0, 2.5, 7.5 and 15 kg I ha^{-1}), the iodine concentration in the edible plant parts increased when trial plots were fertilized with KI or KIO_3 directly before planting or sowing. The highest iodine accumulation was observed in leafy vegetables (butterhead lettuce, crisp lettuce) followed by tuber vegetables (kohlrabi and radish). In these crops, the desired iodine content [50 - 100 µg I $(100\ g\ FM)^{-1}$] was obtained or nearly achieved at a fertilizer rate of 7.5 IO_3^--I ha^{-1} without a significant yield reduction or degradation in the marketable quality. In contrast, the KI supply at the same rate resulted in much lower iodine enrichment and a yield reduction of > 10 % in some crops. The soil pH significantly affected the iodine biofortification of butterhead lettuce. However, the observed increase in the iodine content under acidic soil conditions (pH 4.5) was most probably due to impaired plant growth. Compared to the before mentioned vegetable species, root vegetables (carrots) and onions showed a weak iodine accumulation potential in their edible plant parts. After fertilizing KI or KIO_3, the iodine applied was rapidly converted to non-phytoavailable iodine forms; concordant with this finding, long-term effects of a one-time iodine soil fertilization could not be observed. A comparison between the soil and the foliar fertilization techniques revealed a superior performance of iodine aerially applied on butterhead lettuce. Therefore, this approach should be investigated in further detail, especially on leafy vegetables.

4.2 Introduction

Numerous studies have been conducted to test the agronomical biofortification of iodine on plants by means of soil fertilization. Most trials were carried out as pot experiments in greenhouses under controlled ambient conditions (Borst Pauwels 1961; Muramatsu et al. 1989; Dai et al. 2004a; Hong et al. 2008a, 2009). The investigations showed characteristic response patterns depending on the oxidation state of the element (I^-/IO_3^-) as well as the applied iodine dose and the examined plant species. Concentrations in the range of 5 - 25 mg I (kg soil)$^{-1}$ (≈ 15 - 75 kg I ha^{-1} in the 0 - 30 cm soil layer) without yield impairment have been reported (Dai et al. 2004a; Hong et al. 2008a). Our own preliminary pot trials with butterhead lettuce cultivated in peat growing media showed however, that concentrations in the range of 1 - 2.5 mg I (L substrate)$^{-1}$ (≈ 3 - 7.5 kg I ha^{-1} in the 0 - 30 cm soil layer) were sufficient to achieve a reasonable iodine accumulation in edible plant parts. Concentrations of ≥ 10 mg I (kg substrate)$^{-1}$ (≈ 30 kg I ha^{-1} in the 0 - 30 cm soil layer) induced yield depression (unpublished results).

Only a few trials had been conducted under field conditions. Exogenous factors may have manifold influence on the iodine accumulation behavior of different plant species. For example, it has been suggested that soil pH and organic matter content play important roles in the iodine dynamics of soils (Whitehead 1975; Steinberg et al. 2008). The influence of ubiquitous bacteria, e.g. *Rhizobium* sp. or *Variovorax* sp. may also be of importance for the methylation of iodine in terrestrial ecosystems (Amachi et al. 2001). In addition, soil acidity is known to constrain bacteriological activity (Graham 1981; Munns and Franco 1982; Munns 1986), e.g. limiting *Rhizobium* survival and persistence (Graham et al. 1982; Hartel and Alexander 1983; Brockwell et al. 1991). However, until now no field trials had been run to test the influence of different soil pH-levels on the iodine uptake by plants.

The large scale application of mineral iodine fertilizers occurred, so far, only as drip irrigation or by means of single applications into the irrigation water in areas with severe iodine deficiency (DeLong et al. 1997; Ren et al. 2008). These methods require large iodine quantities and imply an uncontrolled iodine release into the environment. The more precise and cost effective application of iodine by means of spreaders and granulated compound fertilizers has been established in the field of livestock production (Grasstrac®; Yara 2014a). Furthermore, the effects of a one-time iodine soil fertilization over time have been poorly investigated to date and practical recommendations for the use of straight or compound iodine fertilizers on vegetables, to enhance the iodine uptake of food crops to an adequate extent, have not yet been explored.

Foliar sprays are known to be an efficient substitute to soil fertilization, especially in the case of micronutrient (MARSCHNER 2012). Nevertheless, only a few studies have hitherto been conducted to compare the two iodine application methods and the partially inconsistent results justify further investigations in this regard (SMOLEŃ ET AL. 2011a, 2011b). Our own preliminary foliar application trials on kohlrabi in greenhouses showed that a dose of 1 kg I ha^{-1}, especially sprayed as KIO_3, did not negatively affect the leaf surface, whereas doses of 2.5, 7.5 and 15 kg I ha^{-1} caused severe leaf burn (unpublished results).

Therefore, the main objectives of this section were to conduct trials under open field conditions in order to:

- Investigate the long-term effects of a one-time iodine fertilization on crop yield and iodine accumulation in vegetables based on the example of butterhead lettuce, kohlrabi and radish (leafy vegetables vs. stem tuber vegetables).

- Compare the iodine accumulation behavior of a selection of different field vegetable species.

- Determine the influence of different soil pH-levels on the iodine uptake of vegetables using the example of butterhead lettuce.

- Evaluate the efficiency of soil versus foliar fertilization techniques using the example of butterhead lettuce and kohlrabi (leafy vegetables vs. stem tuber vegetables).

4.3 Experimental setup

The soil fertilization experiments were carried out in 2010 and 2011 on sandy loam soils (Sl_3 and Sl_4) at two different sites of the horticultural research station of the University of Applied Sciences, Osnabrück, Germany. Field trials with a focus on soil fertilization over time (from April 2010 until June 2011) and the comparison of soil fertilization versus foliar spray were conducted on arable fields at the site Wulveskamp (N 52° 18' 41.299" - E 8° 1' 30.31"). Soil sampling took place just before iodine fertilization in April 2010 and at intervals of approximately six months. To investigate the short-term dynamics of iodine a further soil application experiment with KIO_3 was conducted and sampling took place after 1 day and 1, 2, 4 and 8 weeks after fertilization. The sampling techniques are described thoroughly in section 3.3. Experiments with a focus on soil pH and different vegetable species were carried out on arable fields at the site Haste (N 52° 18' 20.392" - E 8° 2' 20.28"). A trial overview and the experimental setup are given in Table 4.1. The application technique for iodine salts is described in section 3.2. Briefly, plots were drenched with potassium iodide or potassium iodate solutions at different concentrations one day before planting or sowing at a rate of approximately 2 L H_2O m^{-2}. Foliar sprays were applied 1 and 2 weeks before harvest at different concentrations at a rate of 600 L H_2O ha^{-1}.

Plant material was purchased at Jungpflanzen Lüske GbR, Höltinghausen, Germany. Kohlrabi seedlings (*Brassica oleracea* L. var. *gongylodes* L. `Lech´), butterhead lettuce (*Lactuca sativa* L. var. *capitata* cv. `Barilla´), crisp lettuce (*Lactuca sativa* L. var. crispa cv. `Robinson´) and onion (*Allium cepa* L. agg. *cepa* `Hytech´) all grown in peat substrate were transplanted into soil within two days of delivery (transport and storage temperature: 4 °C). Carrot (*Daucus carota* L. subsp. *sativus* (Hoffm.) Arcang. cv. `Merida´ $F_1$) and radish (*Raphanus sativus* L. var. *sativus* cv. `Raxe´) precision seeds (Hild Samen GmbH, Marburg, Germany) were sowed with a single-seed precision hand-pushed seed drill (Sembdner Maschinenbau GmbH, Fürstenfeldbruck, Germany) at a density of 160 kernel m^{-2}.

The basic N, P and K fertilization was conducted manually, three days before planting or sowing, by spreading granular fertilizers at the amounts shown in Table 4.2. Nitrogen fertilization occurred according to the N_{min}-method (Wehrmann and Scharpf 1986) using specified

N-values for the respective vegetables (FELLER ET AL. 2007) and Entec® 26 (7.5 % NO_3^--N, 18.5 % NH_4^+-N, 13 % S; Compo GmbH & Co. KG, Münster, Germany) as nitrogen fertilizer. Superphosphate (18 % P_2O_5) and potassium magnesia (30 % K_2O, 10 % MgO 18 % S; both K+S AG, Kassel, Germany) were used to cover the P, K and Mg requirements. These were ascertained using the VDLUFA (1997) method for the replacement of nutrients consumed by field crops according to classification into nutrient levels (ALT AND WIEMANN 1987; KERSCHBERGER ET AL. 1997; KERSCHBERGER AND FRANKE 2001).

Climatic data was collected and monitored within the horticultural research station at the Haste site. An overview of climatic conditions during the cultivation period is given in Figure 4.1. Briefly, the Osnabrück region, located in south-west Lower-Saxony, is characterized by a warm-moderate climate with mild winters and cool summers. The long-term averages for minimum and maximum air temperature, rainfall and rain days are 1.8 °C (January), 17.6 °C (July), 865 mm and 122 d, respectively (DWD 2013).

The standard analytical methods as well as the analytical procedures for iodine determination and the statistical procedures used are described in detail in chapter 3.

Table 4.1 Overview of the trial setup and objectives arranged by experimental groups

Exp. no.	Trial year	Trial objectives	Treatments	Cultivated vegetable species	Trial design
1	2010 2011	Test the influence of a one-time iodine soil fertilization over two years using the example of butterhead lettuce and kohlrabi (leafy vegetables vs. stem tuber vegetables)	KI, KIO_3 at 0, 1.0, 2.5, 7.5 and 15 kg I ha^{-1}	2010: **Butterhead lettuce** / **Kohlrabi** Subsequently, after a fallow period without iodine reapplication 2011: **Radish** / **Butterhead lettuce**	Split-plot Repetitions: 3 Gross plot area: 11 m^2
2	2010	Test the iodine accumulation and distribution on different vegetable species after a single iodine fertilization	KI, KIO_3 at 0 and 7.5 kg I ha^{-1}	**Carrot** **Radish** **Onion** **Crisp lettuce** **Kohlrabi**	Randomized complete block design Repetitions: 4 Gross plot area: 9.2 m^2
3	2010	Test the influence of different soil-pH-levels on the iodine uptake of butterhead lettuce	**pH 4.5**, **pH 5.0**, **pH 5.5**, **pH 6.0**, **pH 6.5** combined with KI and KIO_3 at 0 and 7.5 kg I ha^{-1}	**Butterhead lettuce** *Lactuca sativa* var. \`Barilla´	Strip-split-plot Repetitions: 4 Gross plot area: 10 m^2
4	2010	Comparison of soil vs. foliar fertilization techniques on selected crops (leafy vegetables vs. stem tuber vegetables)	**Soil fertilization**: KI and KIO_3 at 0, 1.0, 2.5, 7.5 and 15 kg I ha^{-1} **Foliar sprays:** KI and KIO_3 at 0, 1x 0.5, 2x 0.5, 1x 1.0 and 2x 1.0 kg I ha^{-1}	**Butterhead lettuce** *Lactuca sativa* var. \`Barilla´ **Kohlrabi** *Brassica oleracea* var. *gongylodes* L.	Split-plot Repetitions: 4 Gross plot area: 4.5 m^2 (foliar spray) 11 m^2 (soil fertilization)

Table 4.2 Overview of the growing conditions arranged by experimental group. (*) = the target soil pH-levels in brackets were intended as guidance levels to achieve distinct pH increments

Exp. no.	Vegetable species	Growing conditions						
		Fertilization			Soil properties		Weed management and pest control	Irrigation
		N [kg ha^{-1}]	P_2O_5 [kg ha^{-1}]	K_2O [kg ha^{-1}]	organic matter content [%]	pH-level (*)		
1	Butterhead lettuce (2010)	124	34	181	1.8	6.3	As required: manual weeding, cultivation guard nets, Butisan® (2 l ha^{-1}; 400 l H_2O ha^{-1}), Banvel® M (4 l ha^{-1}; 400 l H_2O ha^{-1}), Basta® (2.5 l ha^{-1}; 600 l H_2O ha^{-1}), Mesurol® (5 kg ha^{-1})	Overhead sprinkle irrigation as required at 4 mm m^{-2} d^{-1}
1	Kohlrabi	204	46	190	1.9	6.7		
1	Radish	110	21	101	1.8	6.3		
1	Butterhead lettuce (2011)	150	34	181	1.9	6.3		
2	Carrot	69	49	318	1.8	6.0		
2	Crisp lettuce	119	34	181				
2	Onion	169	41	164				
2	Radish	89	21	101				
2	Kohlrabi	209	46	190				
3	Butterhead lettuce	124	34	181	1.8	4.2 (4.5)		
					1.5	4.9 (5.0)		
					1.9	5.4 (5.5)		
					1.6	5.7 (6.0)		
					1.3	6.4 (6.5)		
4	Butterhead lettuce	124	34	181	1.8	6.3		
4	Kohlrabi	204	46	190	1.9	6.7		

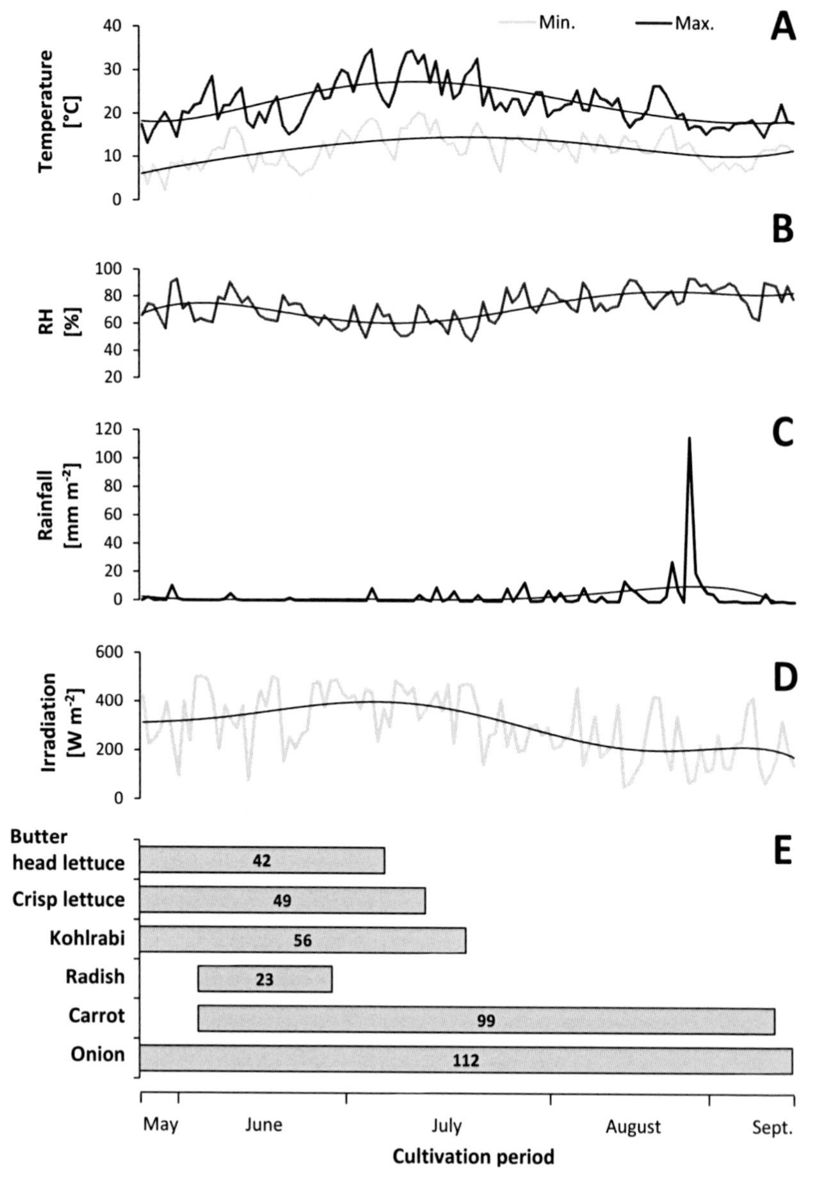

Figure 4.1 Overview of climatic conditions during the cultivation period and respective growth period of different vegetables. Climatic data collected at Osnabrück/Haste meteorological station, Germany, and calculated as daily average. A = maximum and minimum air temperature; B = relative humidity; C = rainfall; D = irradiation; E = total cultivation time in days and relative cultivation time-slot of vegetables investigated in season 2010 (2010.05.25 - 2010.09.13)

4.4 Results

4.4.1 Experiment no. 1: Influence of a one-time iodine soil fertilization over time on selected crops

The influence of a one-time iodine soil fertilization was investigated over two years in experiment no. 1 on leafy and stem tuber vegetables using butterhead lettuce and kohlrabi/radish as model crops. The iodine supply (0 - 15 kg I ha^{-1} applied as KI and KIO_3) did not affect statistically significantly the biomass production of the investigated vegetables in any case (Table 4.3). However, in the first growing season the mean crop yield was reduced by more than 20 % in butterhead lettuce fertilized with both KI and KIO_3 at a dose of 15 kg I ha^{-1} and in kohlrabi at 7.5 and 15 kg I^--I ha^{-1}.

Table 4.3 The influence of different iodine application doses and forms on the crop yield of selected crops. Fresh matter yield is expressed as percent of the unfortified control treatment. Percentages with same letters do not differ according to Bonferroni MCP at $\alpha = 0.05$. Levels of significance are represented by * = $p < 0.05$, ** = $p < 0.01$ and *** = $p < 0.001$ and NS = not significant = $p > 0.05$ (actual probability level)

Cultivated plant species	Kohlrabi		Butterhead lettuce		Butterhead lettuce		Radish	
Cultivation year	2010				2011			
Plot area	A		B		A		B	
Treatment dose [kg I ha^{-1}] / Iodine form [I^- / IO_3^-]	I^-	IO_3^-	I^-	IO_3^-	I^-	IO_3^-	I^-	IO_3^-
	Relative crop yield [%]							
0	100.0 a	100.0 a	100.0 a	100.0 a	100.0 a	100.0 a	100.0 a	100.0 a
1.0	90.2 a	98.6 a	97.9 a	92.0 a	103.1 a	93.9 a	108.6 a	94.0 a
2.5	93.1 a	106.5 a	91.7 a	94.3 a	99.4 a	105.5 a	94.4 a	100.0 a
7.5	79.6 a	100.2 a	85.2 a	94.2 a	113.4 a	102.1 a	85.8 a	97.1 a
15	72.1 a	95.5 a	73.6 a	72.7 a	107.0 a	98.1 a	90.5 a	89.2 a
ONE WAY ANOVA (p-Value)	NS (0.181)		NS (0.389)		NS (0.992)		NS (0.059)	

A confirming visual comparison is given using the example of butterhead lettuce (Figure 4.2). Noteworthy, is the slightly decreasing overall head size with increasing fertilizer dose and the inhomogeneous crop population within the same plot at the highest iodine treatments (Figure 4.2 C and F). Furthermore, butterhead lettuce transplants cultivated at 15 kg I^--I ha^{-1} developed chlorotic leaves with yellow intercostal leaf areas turning increasingly into necrotic spots a few days after planting (Figure 4.3 C). Even though showing growth inhibition, most plantlets recovered within a short period of time.

In most cases however, a moderate iodine supply up to 7.5 kg I ha^{-1}, especially applied as KIO_3, did not affect growth, yield and marketable quality of the investigated crops.

Figure 4.2 A visual comparison of butterhead lettuce one day before harvest, cultivated at different iodine doses: A = 0 kg I^--I ha^{-1}, B = 7.5 kg I^--I ha^{-1}, C = 15 kg I^--I ha^{-1}, D = 0 kg IO_3^--I ha^{-1}, E = 7.5 kg IO_3^--I ha^{-1}, F = 15 kg IO_3^--I ha^{-1}. The white arrows indicate smaller heads within a single plot

Figure 4.3 A visual comparison of butterhead lettuce soil cube transplants eight days after planting at different iodine doses: A = 0 kg I^--I ha^{-1}, B = 7.5 kg I^--I ha^{-1}, C = 15 kg I^--I ha^{-1}, D = 0 kg IO_3^--I ha^{-1}, E = 7.5 kg IO_3^--I ha^{-1}, F = 15 kg IO_3^--I ha^{-1}. The white arrows indicate chlorotic intercostal areas or necrotic spots

Diagram A in Figure 4.4 shows the iodine accumulation behavior of kohlrabi cultivated in the first season after a single iodine soil fertilization applied just before planting in 2010. An increasing iodine content in edible plant parts was observed with augmented iodine supply, particularly when using KIO_3 as the fertilizer. The desirable iodine amount in edible plant parts was achieved at ≥ 7.5 kg IO_3^--I ha^{-1}. KI treatments were remarkably less effective and did not reach the target range [50 - 100 µg I (100 g FM)$^{-1}$] in any trial.

A similar pattern, but at higher accumulation levels, was found in butterhead lettuce cultivated in 2010 (Figure 4.4 B). Again, an increasing iodine content with augmented iodine supply and a higher accumulation tendency of trial variants treated with IO_3^- was observed. Significant differences to the unfortified control occurred at ≥ 7.5 kg IO_3^--I ha^{-1} where the intended iodine level was exceeded to the extent of 50 - 300 %. In I^--treatments at the same iodine doses, a significantly lower iodine accumulation was found; here accomplishing the target range at only 15 kg I^--I ha^{-1}.

In the second season (2011), the rotational crops cultivated on the same plots without further iodine fertilization showed only little or no iodine accumulation (Figure 4.5 A and B). In both succeeding crops, a single significant difference to both unfortified controls was found in plots fertilized with 15 kg IO_3^--I ha^{-1} one year before. Even at the aforesaid dose, the recorded iodine accumulation was distinctly below the desirable range.

The comparison of different datasets, as two or three factorial ANOVA, demonstrates clearly that the factors iodine dose and vegetable species did significantly influence the iodine content in the edible parts of the vegetables species investigated (Table 4.4). The influence of the factor iodine form could not be proven significant in all trials, especially those in season 2011 with a rather poor iodine accumulation in the edible plant parts.

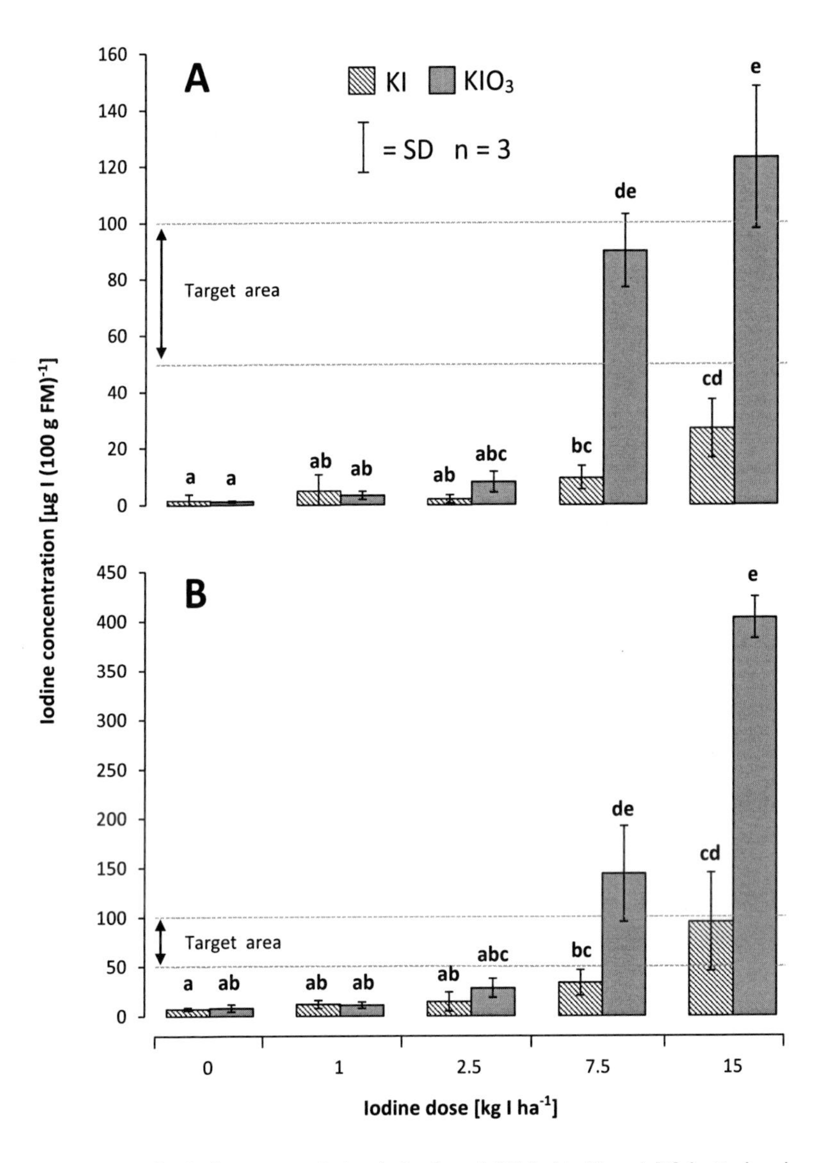

Figure 4.4 The iodine accumulation behavior of [A] kohlrabi and [B] butterhead lettuce cultivated in season 2010 after a one-time iodine soil fertilization applied 2010. Means with same letters do not differ according to Bonferroni MCP at $\alpha = 0.05$ (One-way analysis of variance [A]: probability level = 0.00, power = 1.00; One-way analysis of variance [B]: probability level = 0.00, power = 1.00)

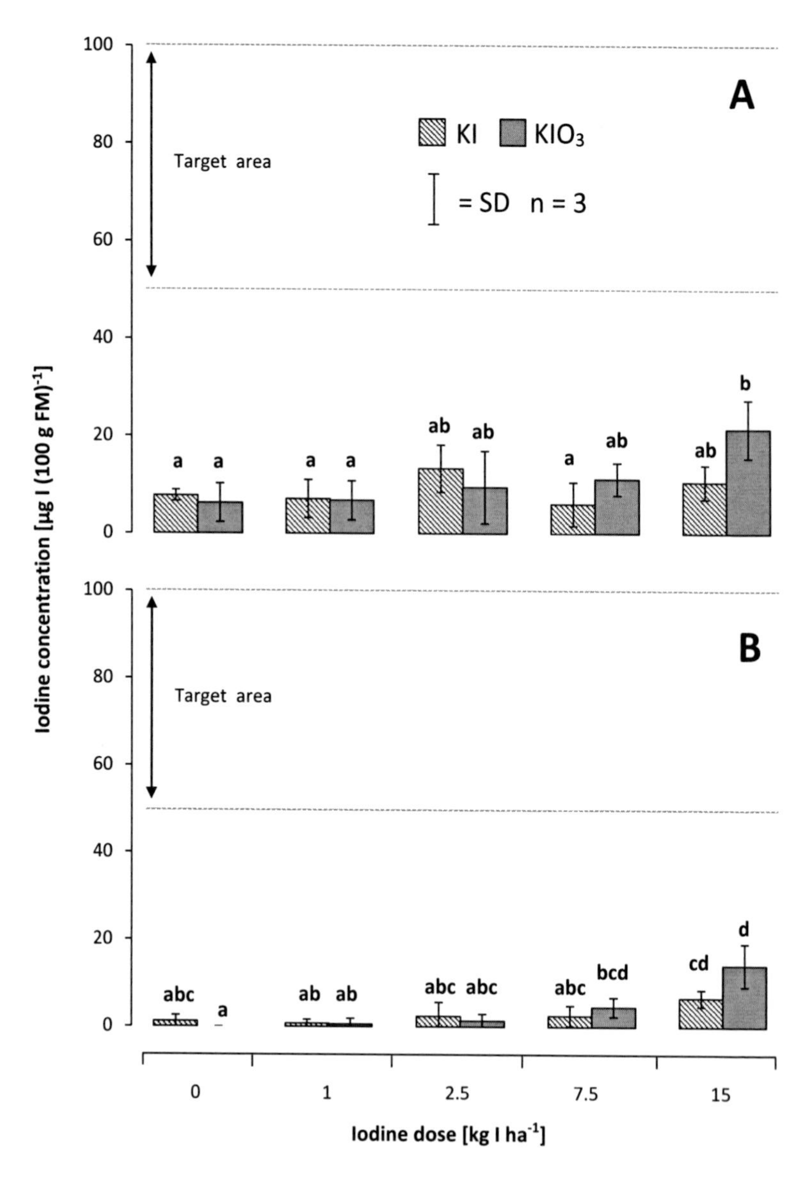

Figure 4.5 The iodine accumulation behavior of [A] butterhead lettuce and [B] radish cultivated in season 2011 after a one-time iodine soil fertilization applied 2010. Means with same letters do not differ according to Bonferroni MCP at α = 0.05 (One-way analysis of variance [A]: probability level = 0.037115, power = 0.805387; One-way analysis of variance [B]: probability level = 0.000012, power = 0.99999)

Table 4.4 Analysis of variance tables showing the influence of blocks, iodine dose, iodine form and their respective interactions as affected by the different iodine application doses and forms on the iodine accumulation behavior of selected crops. Levels of significance are represented by * = p < 0.05, ** = p < 0.01, *** = p < 0.001 and NS = not significant = p > 0.05 (actual probability level)

Cultivated plant species		Kohlrabi	Butterhead lettuce	Butterhead lettuce	Radish
Cultivation year		**2010**		**2011**	
Plot area		**A**	**B**	**A**	**B**
GLM ANOVA two factorial	Block	NS (0.694226)	NS (0.065582)	NS (0.878126)	NS (0.136980)
	Iodine form (F)	* (0.040881)	NS (0.086514)	NS (0.495168)	NS (0.768978)
	Iodine dose (D)	*** (0.000000)	*** (0.000000)	* (0.038576)	*** (0.000001)
	F x D	*** (0.000965)	** (0.002178)	NS (0.116519)	* (0.046248)
GLM ANOVA three factorial between species within season	Block	NS (0.132904)		NS (0.575901)	
	Species (S)	** (0.004787)		** (0.001671)	
	Iodine form (F)	*** (0.000000)		NS (0.150624)	
	Iodine dose (D)	*** (0.000000)		*** (0.000001)	
	S x F	NS (0.330249)		NS (0.677177)	
	S x D	NS (0.460005)		NS (0.163035)	
	F x D	*** (0.000002)		** (0.007015)	
	S x F x D	NS (0.761078)		NS (0.898697)	
GLM ANOVA three factorial between species over seasons A	Block	NS (0.636253)			
	Species (S)	** (0.002653)			
	Iodine form (F)	*** (0.000000)			
	Iodine dose (D)	*** (0.000000)			
	S x F	* (0.015900)			
	S x D	*** (0.000000)			
	F x D	*** (0.000000)			
	S x F x D	*** (0.000106)			
GLM ANOVA three factorial between species over seasons B	Block		* (0.018291)		
	Species (S)		*** (0.000024)		
	Iodine form (F)		*** (0.000661)		
	Iodine dose (D)		*** (0.000000)		
	S x F		NS (0.052566)		
	S x D		** (0.004963)		
	F x D		*** (0.000488)		
	S x F x D		NS (0.267222)		

Confirming the results gained from crops cultivated in the second season, only modest differences were registered in the iodine content of soil samples collected before and six months after iodine fertilization (Figure 4.6 A). Although a slight increase in iodine content in the soil was observed with increasing iodine fertilizer doses, especially in soil depths of 60 - 90 cm in case of KIO_3, no statistically significant differences to the control treatment or to the ambient level value before the fertilization occurred, were computed.

The decrease in iodine recovery following a one-time iodine fertilization (7.5 kg IO_3^--I ha^{-1}) at shorter sample collection intervals is remarkable (Figure 4.7 B). In this case, a statistically significant reduction in iodine concentration, especially pronounced in the upper soil layer (0 - 15 cm), was found with elapsing time. A rapid decrease was recorded within the first week and already three weeks after the initial application, the majority of the exogenously applied iodine was no longer detectable in the top soil (extracted with a 12.5 mM calcium-chloride solution) without indications of iodine displacement in the deeper soil layer (15 - 30 cm).

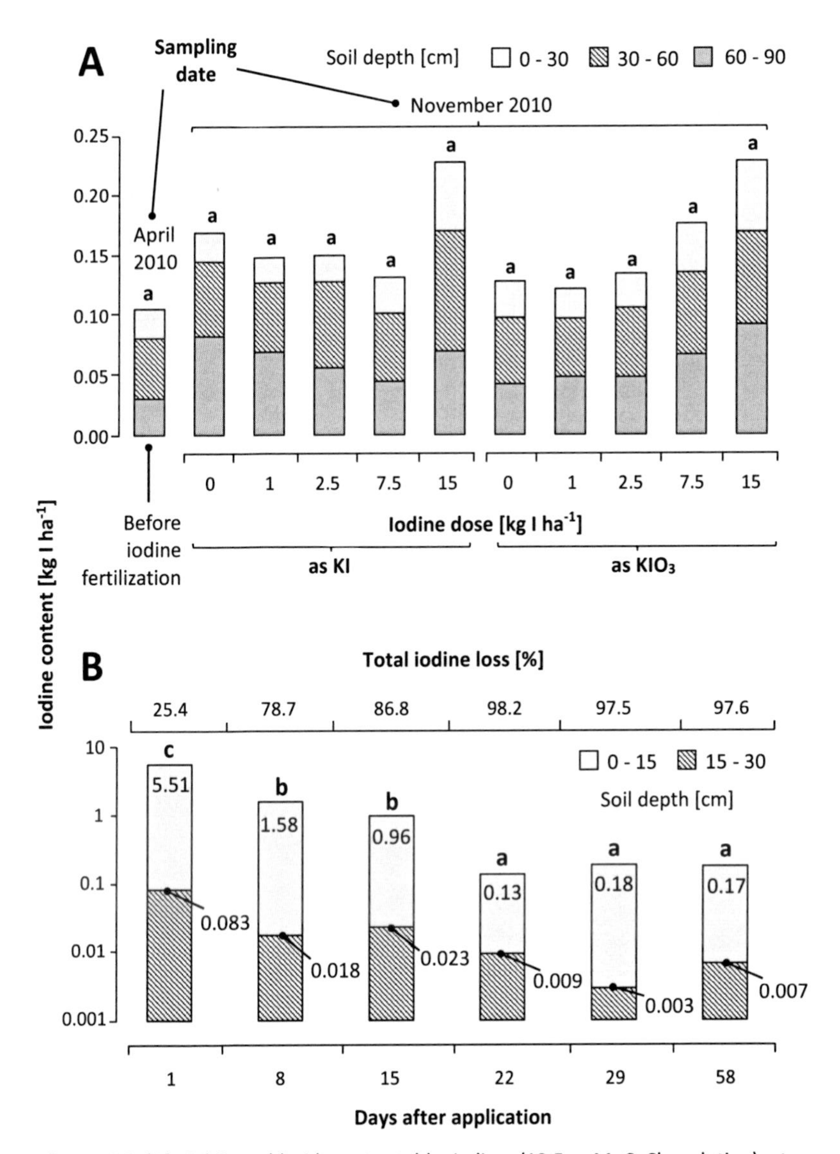

Figure 4.6 [A] Calcium-chloride extractable iodine (12.5 mM $CaCl_2$ solution) at different depths in a sandy loam soil (Sl_3 - Sl_4) before and six months after iodine fertilization by means of soil drenches at different concentrations. Means with same letters do not differ according to Bonferroni MCP at $\alpha = 0.05$ (One-way analysis of variance: probability level = 0.053, power = 0.78). n = 3. [B] Calcium-chloride extractable iodine in soil samples collected at different intervals after a one-time iodine fertilization at 7.5 kg IO_3^--I ha^{-1} (One-way analysis of variance: probability level = 0.00, power = 1.00). n = 4

4.4.2 Experiment no. 2: Iodine accumulation behavior of different vegetable species

In experiment no. 2 five different field vegetable species (including leafy, stem tuber and root crops) were tested for their iodine accumulation behavior using a single soil fertilization of 7.5 kg I ha^{-1} applied as KI and KIO_3. Although compared to the respective control treatment a slight biomass promotion was observed in the carrot and kohlrabi and an extenuated yield tendency in radish and crisp lettuce, no statistically significant differences in crop yield were found (Figure 4.7 A).

The iodine accumulation in edible plant parts shows statistically significant differences between both the vegetable species and the different iodine fertilization treatments (Figure 4.7 B). As previously observed, the iodate treatments were generally more effective than iodide. Noteworthy are the general differences in the iodine accumulation behaviour among species, which clusters radish, kohlrabi and crisp lettuce into an accumulation level group close to the desirable iodine amount in edible plant parts. Onions and carrots were the species with the least observable iodine accumulation.

The iodine content in different plant organs and its relative distribution within the plant are reported in Table 4.5 and Figure 4.8, respectively. Onions were excluded from partition because of leaf absence at harvest time. Once more, iodate treatments showed the highest accumulation levels regardless of plant organ or vegetable species. Generally, the iodine distribution pattern was coherent with the total quantity of fresh matter of a single plant. The higher fresh masses of kohlrabi and crisp lettuce showed a pronounced total iodine accumulation, especially in edible plant parts. The above-ground plant parts account for a higher share of the total iodine content per plant. In each case, the iodine distribution patterns are very different depending on the vegetable species analysed and the chosen plant organ.

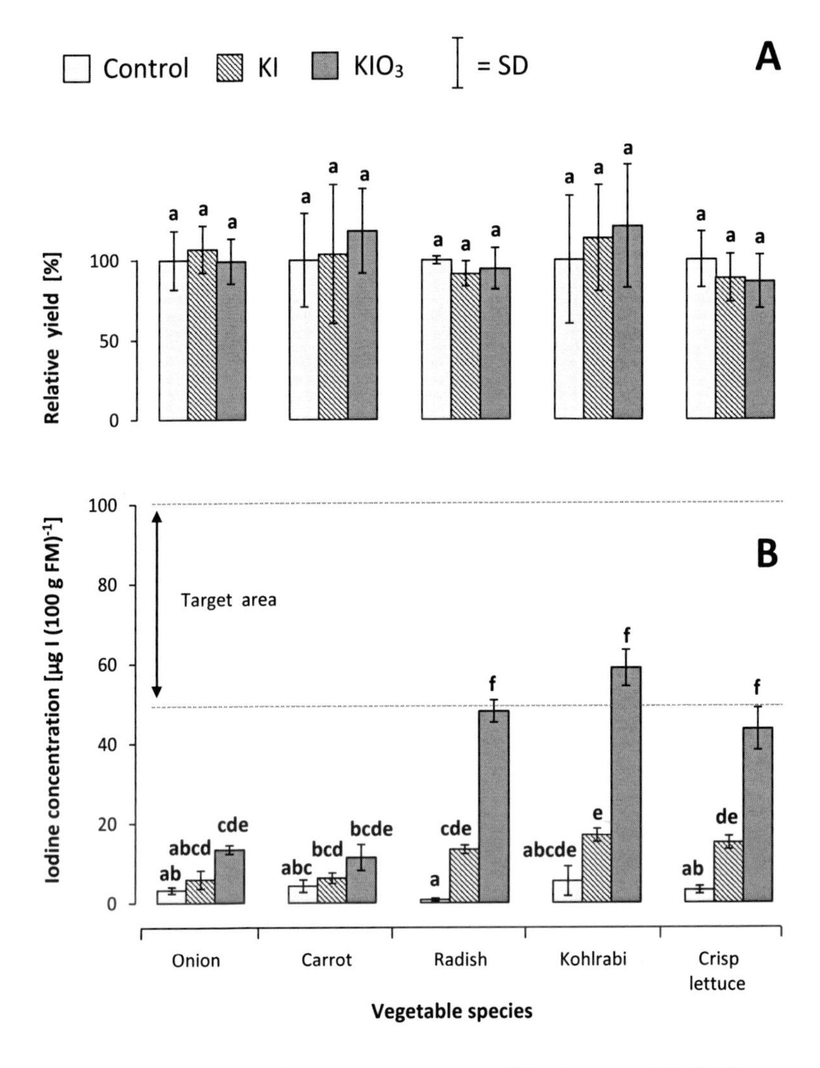

Figure 4.7 Relative yield [A] and iodine content in edible plant parts [B] of different vegetable species as affected by iodine fertilization at 0 kg I ha^{-1}, 7.5 kg I^--I ha^{-1} and 7.5 kg IO_3^--I ha^{-1}. Means with same letters do not differ according to Bonferroni MCP at α = 0.05 (One-way analysis of variance [A]: p-level = 0.4313, power = 0.545287; One-way analysis of variance [B]: p-level = 0.00, power = 1.00; GLM ANOVA [B]: iodine fertilization (I) p-level = 0.00, power = 1.00, vegetable species (S) p-level = 0.00, power = 1.00, interaction I x S p-level = 0.00, power = 1.00). n = 4

Table 4.5 The iodine concentration in different plant organs of a vegetable species selection as affected by iodine fertilization at 0 kg I ha^{-1}, 7.5 kg I^--I ha^{-1} and 7.5 kg IO_3^--I ha^{-1}. Data was transformed into ranks and compared to respective control by the Wilcoxon rank-sum test. Levels of significance are represented by * = $p < 0.05$ and NS = not significant. n = 4

Cultivated plant species	Plant organs	Treatments		
		0 kg I ha^{-1}	7.5 kg I^--I ha^{-1}	7.5 kg IO_3^--I ha^{-1}
		Iodine concentration [µg I (100 g FM)$^{-1}$]		
Carrot	Foliage	11.0	42.0 *	83.6 *
	Root	4.2	6.2 NS	11.4 NS
Radish	Foliage	18.5	80.1 *	305.2 *
	Stem tuber	0.8	13.4 *	48.1 *
	Root	7.7	43.2 *	153.7 *
Kohlrabi	Leaf blade	6.6	23.0 *	81.2 *
	Leaf stalk	3.9	36.1 *	136.0 *
	Stem tuber	5.4	17.0 NS	58.9 *
Crisp lettuce	Outer leaves	10.5	21.8 *	53.1 *
	Inner leaves	9.4	15.2 *	28.7 *
	Stalk	9.6	27.0 *	57.6 *

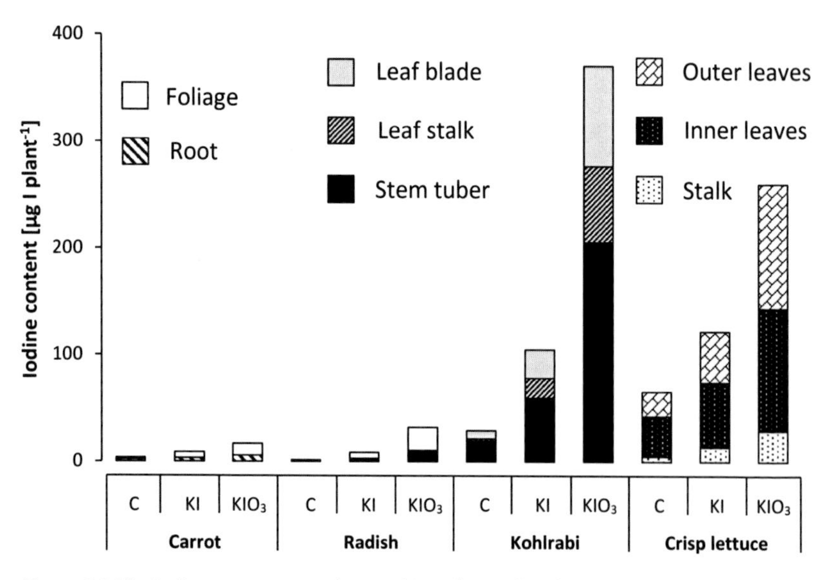

Figure 4.8 The iodine content per plant and its relative distribution in different vegetable species as affected by iodine fertilization at 0 kg I ha^{-1} (= C), 7.5 kg I^--I ha^{-1} (= KI) and 7.5 kg IO_3^--I ha^{-1} (= KIO_3). n = 4

4.4.3 Experiment no. 3: Influence of different soil pH-levels on the iodine accumulation behavior of butterhead lettuce

The influence of soil pH variations on the iodine uptake of butterhead lettuce was tested in experiment no. 3 using a single soil fertilization of 7.5 kg I ha^{-1} applied as KI and KIO_3. Diagram A in Figure 4.9 shows an increasing crop yield tendency with increasing soil pH levels. The iodine treatment means at the lowest soil pH (4.5) were significantly different compared to the means at the highest pH values (5.5 - 6.5). The different iodine forms within a single soil pH-level group did not affect the biomass production as confirmed by the GLM ANOVA in Table 4.6.

Regarding the dry matter of butterhead lettuce, an inverse pattern was found showing decreasing dry matter content with increasing soil pH-level (Figure 4.9 B). Hence, the pH-level group 4.5 produced the highest dry matter percentage which was significantly different from the highest pH-level treatments. Again, the factor pH-level shows significant differences, whereas the iodine form does not (Table 4.6).

A general diminishing tendency for the iodine concentration (in 100 g of edible plant parts) was observed with an increasing soil pH-level (Figure 4.10). Here, significant differences between both iodine forms and the pH-levels were recorded (Table 4.6). Again, the highest iodine contents were observable when using KIO_3 as the iodine fertilizer. The targeted range was almost met or exceeded in all iodine treatments, even when KI was applied.

The iodine content per plant (Figure 4.10 top X-axis) revealed a distribution pattern inverse to the iodine concentration, i.e. an augmented iodine content with increasing pH-level was observed. Yet again, significant differences in iodine form, pH level and their interaction were measurable (Table 4.6).

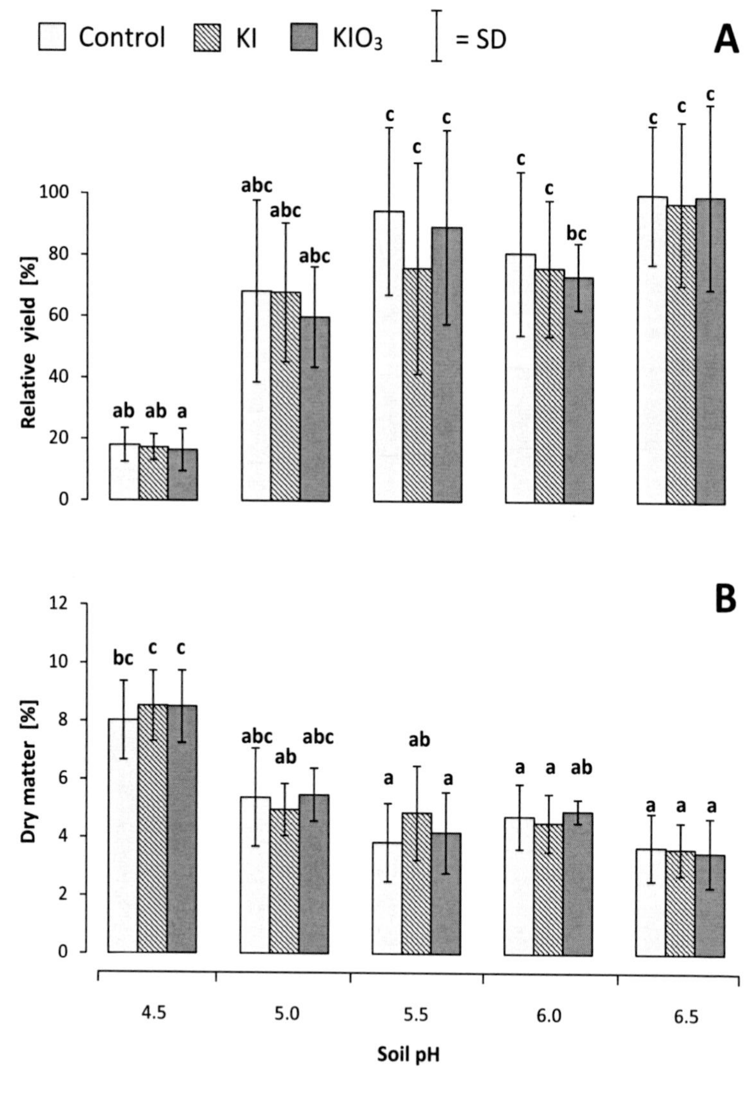

Figure 4.9 Relative yield [A] and dry matter [B] of butterhead lettuce as affected by iodine fertilization (0 kg I ha^{-1}, 7.5 kg I^{-}-I ha^{-1} and 7.5 kg IO_3^{-}-I ha^{-1}) at different soil pH-levels. Means with same letters do not differ according to Bonferroni MCP at α = 0.05 (One-way analysis of variance [A]: probability level = 0.00, power = 1.00; One-way analysis of variance [B]: probability level = 0.00, power = 1.00). n = 4

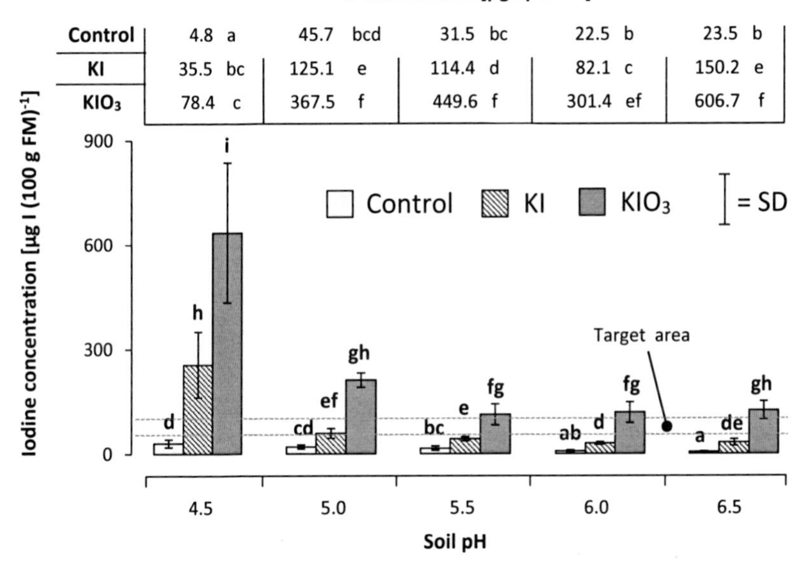

Figure 4.10 Iodine concentration in edible plant parts of butterhead lettuce as affected by iodine fertilization (0 kg I ha^{-1}, 7.5 kg I^{-}-I ha^{-1} and 7.5 kg IO_3^{-}-I ha^{-1}) at different soil pH-levels. Iodine content per plant calculated on basis of the fresh mass yield (per plant). Means with same letters do not differ according to Bonferroni MCP at $\alpha = 0.05$ (One-way ANOVA of total iodine: probability level = 0.00, power = 1.00; One-way ANOVA of relative iodine content: probability level = 0.00, power = 1.00). n = 4

Table 4.6 Analysis of variance table of crop yield, dry matter and iodine accumulation (relative and total) shown as GLM ANOVA. Levels of significance are represented by * = p < 0.05, ** = p < 0.01, *** = p < 0.001 and NS = not significant = p > 0.05 (actual probability level)

Parameter	GLM ANOVA	Probability level	
Crop yield	Block	* (0.0036)	
	Iodine form (F)	NS (0.8191)	
	pH-level (P)	*** (0.0002)	
	F x P	NS (0.6040)	
Dry matter	Block	** (0.004)	
	Iodine form (F)	NS (0.9327)	
	pH-level (P)	*** (0.0001)	
	F x P	NS (0.3389)	
Iodine accumulation (per 100 g FM | per plant)	Block	NS (0.3261)	NS (0.164956)
	Iodine form (F)	*** (0.000001)	*** (0.000000)
	pH-level (P)	*** (0.000000)	*** (0.000013)
	F x P	*** (0.00014)	* (0.010913)

4.4.4 Experiment no. 4: Comparison of soil vs. foliar fertilization technique on two model crops

In experiment no. 4 datasets of two soil fertilization trials (0 - 15 kg I ha^{-1} applied as KI and KIO_3) were compared to the datasets of two parallel foliar application trials to investigate possible differences between the application methods. In the latter case, KI and KIO_3 were applied once or twice at a rate of 0.5 and 1.0 kg I ha^{-1}. For this method comparison butter head lettuce and kohlrabi were chosen to represent leafy and stem tuber vegetables. The biomass production was not affected in either case significantly by the iodine supply (Table 4.7).

Table 4.7 Comparison of iodine application methods on the crop yield of kohlrabi and butterhead lettuce. Yield expressed as percent of control treatment. Percentages with same letters do not differ. Levels of significance are represented by * = $p < 0.05$, ** = $p < 0.01$, *** = $p < 0.001$ and NS = not significant = $p > 0.05$ (actual probability level)

Cultivated plant species		Kohlrabi				Butterhead lettuce			
Application method		Soil drenches		Foliar sprays		Soil drenches		Foliar sprays	
Treatment dose [kg I ha^{-1}]		Iodine form [I^- / IO_3^-]							
Soil drenches	Foliar sprays	I^-	IO_3^-	I^-	IO_3^-	I^-	IO_3^-	I^-	IO_3^-
		Relative crop yield [%]							
0	**0**	100.0 a	100.0 a	100.0 a	100.0 a	100.0 a	100.0 a	100.0 a	100.0 a
1.0	**0.5**	90.2 a	98.6 a	100.6 a	93.5 a	97.9 a	92.0 a	98.3 a	107.4 a
2.5	**2x 0.5**	93.1 a	106.5 a	104.7 a	98.6 a	91.7 a	94.3 a	89.5 a	99.2 a
7.5	**1.0**	79.6 a	100.2 a	99.7 a	97.4 a	85.2 a	94.2 a	94.9 a	99.2 a
15	**2x 1.0**	72.1 a	95.5 a	98.0 a	95.8 a	73.6 a	72.7 a	86.8 a	91.9 a
ONE WAY ANOVA (p-Value)		NS (0.181)		NS (0.875)		NS (0.389)		NS (0.099)	

A visual comparison of the foliar sprays applied to butterhead lettuce illustrates no noticeable differences from the unfortified control in the overall head size with increasing fertilizer dose (Figure 4.11). Similar results were achieved spraying kohlrabi at the same iodine doses). Hence, in contrast to soil drenches, the iodine foliar sprays did not affect in any case the biomass production and the marketable quality of the crops investigated.

Figure 4.11 A Visual comparison of butterhead lettuce one day before harvest, sprayed at different iodine doses: A = 0 kg I^--I ha^{-1}, B = 1 kg I^--I ha^{-1}, C = 2x 1 kg I^--I ha^{-1}, D = 0 kg IO_3^--I ha^{-1}, E = 1 kg IO_3^--I ha^{-1}, F = 2x 1 kg IO_3^--I ha^{-1}

Table 4.8 displays the iodine accumulation behavior of kohlrabi and butterhead lettuce as affected by iodine foliar sprays and the direct comparison to the respective soil application trial. Generally, the foliar sprays applied to kohlrabi did not lead to a remarkable iodine accumulation in edible plant parts. In all cases, this was far below from the desired minimal amount of 50 µg I (100 g FM)$^{-1}$. Non-parametric comparisons to the control level and between the application methods showed a slight iodine enhancement when applying KI at the highest dose and a superior performance of iodate in soil application on kohlrabi. Foliar spray treatments on butterhead lettuce led to very high iodine accumulation levels exceeding the soil application results. In contrast to the soil drenches, a more pronounced iodine accumulation tendency was observed in the iodide treatments.

Remarkably, the targeted accumulation range could be reached already at the lowest doses of 0.5 kg I^--I ha^{-1} and 2x 0.5 kg IO_3^--I ha^{-1}. Significant differences were measured regarding the factors iodine form and dose, as well as for the comparison between the application methods on butterhead lettuce and between the two vegetable species within the same application method (ANOVA and parameter free tests in Table 4.8).

Table 4.8 Comparison between the application methods on the iodine accumulation behavior of kohlrabi and butterhead lettuce. Means represent the iodine concentration per 100 g of edible plant parts. Analysis of variance and nonparametric tests table: Levels of significance are represented by * = $p < 0.05$, ** = $p < 0.01$, *** = $p < 0.001$ and NS = not significant = $p > 0.05$ (actual probability level). Means with same letters do not differ according to Bonferroni MCP at $\alpha = 0.05$. Kohlrabi foliar spray data was transformed into ranks and compared to respective control by Wilcoxon rank-sum test. Butterhead lettuce application method comparison is shown as a parameter free method since the pooling of datasets leads to non-normal data. Soil drenches n = 3. Foliar sprays n = 4

Cultivated plant species 〉		Kohlrabi				Butterhead lettuce			
Application method 〉		Soil drenches		Foliar sprays		Soil drenches		Foliar sprays	
Treatment dose [kg I ha^{-1}]		Iodine form [I^- / IO_3^-]							
Soil drenches	Foliar sprays	I^-	IO_3^-	I^-	IO_3^-	I^-	IO_3^-	I^-	IO_3^-
		Iodine concentration [µg I (100 g FM)$^{-1}$]							
0	0	1.5 a	22.1 bc	1.3	16.3	7.1 a	8.1 ab	5.3 a	4.8 a
1.0	0.5	4.9 a	29.3 bc	2.1 NS	12.7 NS	11.7 ab	10.9 ab	82 ab	36.5 a
2.5	2x 0.5	1.9 a	41.1 c	2.3 NS	19.8 NS	14.3 ab	27.9 abc	236 bcd	94.9 abc
7.5	1.0	9.4 ab	89.9 d	2.8 NS	18.6 NS	33.3 bc	143.6 de	251.4 cd	147 abcd
15	2x 1.0	26.8 bc	122.9 d	15.1 *	16.5 NS	95.1 cd	403.3 e	473.9 e	283.5 d
ONE WAY ANOVA (Probability value)		see Figure 4.4				see Figure 4.4		*** (0.000000)	
GLM ANOVA two factorial	Block	see Table 4.4				see Table 4.4		NS (0.491921)	
	Iodine form							* (0.025436)	
	Iodine dose							*** (0.000000)	
	F x D							NS (0.153209)	
Nonparametric tests: between treatments	Friedman's test			*** (0.000818)					
Nonparametric tests: between application methods	Wilcoxon rank-sum test	** (0.009325)				* (0.017902)			
Nonparametric tests: between species	Friedman's test			*** (0.000000)					
	Wilcoxon rank-sum test			*** (0.000000)					

4.5 Discussion

4.5.1 Crop yield and marketable quality as affected by iodine soil fertilization

With regard to crop yield and marketable quality, no statistically significant differences between the trials treatments could be attributed to the iodine application by means of soil fertilization. However, visual comparisons of butterhead lettuce cultivated in 2010 definitely showed a decreasing head size, in particular at the highest iodine dose. The inhomogeneous head sizes within the single plots and the occasional fluctuations in yield between the blocks, also because of the sloped arable field at the Wulveskamp site, probably led to a high data deviation and the consequent non-significant statistical differences in the data.

Nevertheless, a moderate iodine supply of up to 7.5 kg I ha^{-1}, preferably as KIO_3, was found to be well tolerated by the plant species investigated in the soil fertilization trials and is therefore presumably the upper limit for iodine biofortification by means of soil drenches.

Phytotoxicity was observed in the very early developmental stage of butterhead lettuce transplants; where the 15 kg I^--I ha^{-1} dose was applied, this occurred shortly after application. Considering that the highest applied dose of 15 kg I^--I ha^{-1} was calculated for soil depths up to 30 cm [≈ 3.33 mg I^--I (kg soil)$^{-1}$] and soil drenches presumably infiltrated directly after the iodine fertilization only a few centimeters (1.5 - 3 cm) into the soil, the initially higher iodine concentration in the range of approximately 33 - 66 mg I^--I (kg soil)$^{-1}$ could have caused the depicted detrimental effects (TANAKA ET AL. 2012). This would be in concordance with the findings of HONG ET AL. (2008) and WENG ET AL. (2008a) observed on different vegetable species and the postulated deleterious effects of an iodine dose of ≥ 50 mg I^--I (kg soil)$^{-1}$. On the other hand, even though the applied fertilizer dose of 2.5 kg I^--I ha^{-1} covered a range of approximately 5.5 - 11 mg I^--I (kg soil)$^{-1}$ (assuming again soil penetration of 1.5 to 3 cm) a biomass promotion at a low iodine application rate of 10 mg I^--I (kg soil)$^{-1}$, as reported by the same authors, could not be observed.

The visual comparison of butterhead lettuce treated with foliar sprays showed a much more homogenous head size throughout all treatments. The unimpeded crop growth until the application point in time and beyond resulted in well-developed lettuce heads. No adverse effects were noticed on the crop leaves and the pronounced visual differences in biomass production noticed in soil application treatments were not observed in foliar spray trials. With regard to the yield and the external quality of the cultivated crops, the use of foliar sprays was found to have neither negative nor positive influences on the vegetable species investigated. In accordance with these observations, SMOLEŃ ET AL. (2011a, 2011b) found in similar field trials on butterhead lettuce and carrot no significant differences in biomass production. Although statistically not significant, ALTINOK ET AL. (2003) reported yield promotion of alfalfa forage using potassium iodide foliar sprays (1 - 2 kg I^--I ha^{-1}). In contrast, STRZETELSKY ET AL. (2010) stated yield depression spraying radish with KI at 2x 0.8 kg I ha^{-1}. Therefore, the crop yield and the marketable quality of the vegetables, as affected by iodine foliar sprays, are parameters which should be taken in to account in further investigations.

4.5.2 Influence of a one-time iodine soil fertilization over time on different crops

Vegetables cultivated after a one-time iodine soil fertilization at rates of ≥ 7.5 kg IO_3^--I ha^{-1} accumulated iodine in edible plant parts to a satisfactory amount [≥ 50 µg I $(100\ g\ FM)^{-1}$]. In contrast, the vegetables grown on the same plots in the second season without further iodine biofortification did not accumulate iodine to an adequate extent. Thus, long term effects of iodine application by means of soil drenches could not be observed.

The response of vegetables to a one-time iodine soil fertilization was conditioned primarily by the applied **iodine form** and dose, the crop used and the time elapsed. Higher iodine concentrations in vegetables were found throughout in the IO_3^- treatments of the experiments and are consistent with the findings of DAI ET AL. (2006) and WENG ET AL. (2008a) who reported a distinctively higher iodine accumulation fertilizing with the oxidized form IO_3^-.

The inherent difference between the two iodine forms has been explained in several studies as being due to the higher mobility of iodide in soils, which leaches from the root zone more quickly and rapidly volatilizes in the form of organoiodides such as methyliodide

(MURAMATSU ET AL. 1995; FUGE 1996; REDEKER ET AL. 2000; DIMMER ET AL. 2001; JOHNSON ET AL. 2002). WHITEHEAD (1975) observed a greater iodate uptake as compared to iodide in ryegrass and explained the better phytoavailability of IO_3^- being due to the longer residence in soil. It has been suggested that iodide is readily fixed in humus and is thereby made unavailable to the plant. Moreover, IO_3^- is transformed into I^- in the presence of plants (MURAMATSU ET AL. 1983; KATO ET AL. 2013). The slower uptake rate of IO_3^-, which was often observed in nutrient solution experiments, is probably limited by the reduction process (BÖSZÖRMÈNYI AND CSEH 1960, ZHU ET AL. 2003) as well as the heavier molecular weight and the higher valence (MACKOWIAK AND GROSSL 1999). Studies on vegetables grown in hydroponic systems have shown as well that roots absorb I^- at a higher rate than IO_3^- (WHITEHEAD 1973; ZHU ET AL. 2003; BLASCO ET AL. 2008; VOOGT ET AL. 2010). Thus it seems that iodide is more readily plant available in the solution of soilless culture systems but, on the other hand, under field conditions more subject to cumulative losses than IO_3^-. This may explain the deleterious effects of iodide on butterhead lettuce transplants observed only shortly after application of the highest I^- dose (Figure 4.3 C) and the subsequent rapid recovery of the plantlets.

Concerning the required **iodine dose**, butterhead lettuce and kohlrabi showed a remarkable difference in the iodine accumulation behavior: Butterhead lettuce proved to be a very good iodine accumulator since this crop distinctly exceeded the desired iodine concentration in edible plant parts at the highest concentrations. Hence, an amount of approximately 5 kg IO_3^--I ha^{-1} would probably be sufficient to match the target level range. Although kohlrabi had a less accentuated response to the iodine treatments, satisfactory results were achieved at 7.5 kg IO_3^--I ha^{-1}. Differences in the accumulation levels may be explained by the distinguishing morphology of the two crops: According to ZHU ET AL. (2003) iodine is transported within plants mainly by the xylem via the transpiration stream. A large leaf area promotes the total transpiration rate and thus the total accumulation of iodine in the leaves of butterhead lettuce. The habitus of kohlrabi contrasts, with a clear separation between edible plant parts and leaves, whereas the constitution of leafy vegetables and the mass flow underlines an acropetally increasing iodine concentration gradient. Hence, a less pronounced accumulation in the enlarged stem is quite plausible and endorses, as reported in literature, the little phloem mobility of iodine (HERRET ET AL. 1962; BLASCO ET AL. 2008; VOOGT ET AL. 2010).

The soil analyses after the first cultivation season (Figure 4.6 A) revealed a high decrease in plant-available iodine. This explains definitely the low iodine accumulation of crops in the second season and emphasizes the short-term bioavailability of the fertilized mineral iodine in soil. More enlightening was the postponed trial with a soil sample collection at short intervals (Figure 4.6 B). Even fertilizing iodine in its oxidized form, a very quick and significant iodine loss without displacement in the deeper soil layers was observed. Hence, the leaching of iodine seems to be, at least for the tested soils (SI_3 - SI_4), a minor loss pathway and is consistent with observations of WENG ET AL. (2009). Other potential loss pathways may be the volatilization of iodine into organoiodide compounds and the fixation into the soil organic matter as well as iron and aluminum oxides (WHITEHEAD 1978, 1981, 1984). In addition, the oxidizing effect of different bacterial strains and enzymes may be of importance for the immobilization of iodine in soils (MURAMATSU AND AMACHI 2007; SHIMAMOTO ET AL. 2011; SUZUKI ET AL. 2012). Thus, the most feasible explanation for the high iodine losses and/or unavailability is a multiple interaction of the aforementioned factors, which were strongly influenced by the general soil composition and physiochemical properties.

Regardless of the loss/retention pathway, iodate reacted within just a few hours of the application, denoting heavily decreasing calcium-chloride extractable iodine concentrations within the first week of application. This trait matches the growing period of the crops investigated: The faster developing butterhead lettuce spent a higher percentage of its total cultivation period in an iodine rich substrate compared to kohlrabi. Hence, the general morphism, habit and constitution of the vegetable species, as well as the form and dose of iodine applied are the primary aspects to consider. The iodine application point in time in relation to the vegetable growing period has to be recognized as crucial factor, since long term effects through iodine fertilization cannot be expected. To better ascertain and understand the loss/retention pathways of iodine in soil, further studies on the different fractions and dynamics of iodine should be conducted in connection with biofortification approaches.

4.5.3 Iodine uptake and distribution in some vegetable crops

The comparison between crops differing greatly in morphology, habit and length of cultivation showed that some of the vegetable species investigated accumulated significantly more iodine in the harvested plant organs. Kohlrabi, radish and crisp lettuce reached or closely approached the targeted iodine range when IO_3^- was fertilized at a rate of 7.5 kg I ha^{-1} (Figure 4.7 B). This is in accordance with previous findings that indicate the oxidized iodine form as best suited for soil fertilization (Figure 4.4 A and 4.10). Radish and kohlrabi, as stem tuber vegetable species of similar constitution (e.g. comparable leaf area to fresh mass ratio), showed a comparable iodine accumulation. When comparing the leafy vegetables crisp lettuce and butterhead lettuce at the same iodine dose (Figure 4.4 B and Figure 4.10), the latter mentioned crop indicates a distinctively higher iodine accumulation potential. This may once again be explained by the morphology of the related species: Crisp lettuce forms a closed head at a very early developmental stage, in which the inner leaves are subject to less transpiration than the leaves of a looser head of butterhead lettuce. This was confirmed by the differentiated organ analysis which showed distinctively higher iodine contents probably due to higher transpiration rates in the outer leaves of crisp lettuce (Table 4.5). Furthermore, the more fibrous plant parts such as the stalk, leaf stalk and roots showed fairly high iodine accumulation levels. These findings are in accordance with the results of WENG ET AL. (2012) who observed an increased incorporation of iodine in the fibrous tissue of roots and a higher accumulation capacity in leafy plant parts in general, but with variations between species within the same vegetable group.

The lowest iodine concentrations were found in onions and carrots. This can be explained in the first instance by the slower development of plantlets in the sown carrots, the lower transpiration rate of onions (SASTRY ET AL. 1978; BECKER AND FRICKE 1992) and the longer total cultivation period compared to the other crops investigated. Moreover, the initial poor germination of carrots and radish, due to high temperatures, required re-sowing and, consequently, led to a postponed uptake in a period of probably poorer iodine availability in soil. These facts again underline the temporal limitations of the bioavailable iodine in the soil and the consequent influence of the iodine application time in relation to the crop sowing or planting dates as well as the total cultivation period on the iodine accumulation in vegetables. Taking account of the differentiated organ analysis and comparing the two sown vegetable species radish and carrot, the aspect vegetable rooting depth becomes a factor as well. Soil drenches only infiltrated a few cm into soil and thus the iodine concentration in deeper root zones remained low as described before in section 4.5.1. Hence, carrots, with a rooting depth up to 60 cm, probably absorbed iodine only in the very upper region of the edible root and accumulated iodine mostly in the foliage. In contrast, radish accumulated a fairly high relative amount of iodine in all plant organs probably because of the more shallow position of the root system as well as the shorter cultivation period and the higher transpiration rate in comparison to carrots (VERMA ET AL. 2007).

4.5.4 Iodine accumulation in butterhead lettuce as affected by soil pH-level

A high influence of the soil pH conditions was observed on the iodine uptake of butterhead lettuce. A clear enhancement of the biomass production was recorded with increasing soil pH and, in turn, a decrease of the dry matter fraction with increasing pH (Figure 4.9 A and B). The relative iodine concentration [µg I $(100\ g\ FM)^{-1}$] increased with decreasing pH but the total iodine content per plant showed an inverse iodine accumulation pattern (Figure 4.10).

One reason for the pronounced iodine concentration at pH 4.5 might be the reaction of iodate with the soil organic matter. This reduction process increases with decreasing pH, delivering HIO or I_2 as intermediates which are further reduced to I^- and released into the soil solution (STEINBERG ET AL. 2008). On the other hand, this effect may have been (partly) neutralized by a higher iodine fixation onto Fe or Al-oxides at low soil pH values. Concordantly, SHEPPARD ET AL. (1997) showed that the soil pH level was not a clear indicator for iodine sorption, being a complex and site-specific parameter heavily dependent on the organic soil matter, the microbial and the enzymatic system and the mineral soil fractions (especially Fe, Mn and Al oxides).

A constrained bacterial activity at low pH-levels and thus a reduced methylation of iodine into volatile organoiodine compounds should be taken into account as well (GRAHAM 1981; BROCKWELL ET AL. 1991; AMACHI ET AL. 2001). In contrast hereto, VARGAS AND GRAHAM (1988), GRAHAM ET AL. (1994) and TIWARI ET AL. (1996) detected acid tolerant *Rhizobium* sp. strains. MIWA ET AL. (2008) isolated the boron accumulating *Variovorax boronicumulans* sp. and observed a growth range of pH 5 - 9. SATOLA ET AL. (2013) discovered that the *Variovorax* sp. isolate WP1 had a pH operating range of 6.5 to below 4 under laboratory conditions. Hence, a low pH may have inhibited the iodine methylation to some extent, but at least some of the ubiquitous bacteria responsible for iodine methylation in soils may be resistant to acidic soil conditions.

A more feasible explanation for the pronounced iodine accumulation at low soil pH involves the findings of the biomass production and the dry matter fraction: Butterhead lettuce cultivated under acidic conditions developed only a small percentage of the typical biomass that can be developed under normal, slightly acidic to neutral growing conditions. The pH induced growth inhibition has probably led to a proportionally larger iodine amount in smaller plant cells. Furthermore, this is supported by the more than doubled dry mass fraction found at pH 4.5. Taking the biomass differences into account and calculating the total iodine content per plant, the picture of an iodine content accretion with increasing pH emerges; i.e. the larger the lettuce head the more total iodine will ultimately be accumulated.

In conclusion, although a clear association of the responsible factors cannot be made at this stage, a great deal may be apportioned to a constrained plant growth as a consequence of a low soil pH. However, a multiple influence of the aspects discussed above cannot be categorically excluded and further trials should be undertaken to clarify the remaining uncertainties.

4.5.5 Efficiency comparison of soil versus foliar fertilization technique

The different application techniques - soil vs. foliar fertilization - showed an intrinsic difference in the iodine accumulation behavior depending on the vegetable species studied. Butterhead lettuce showed higher accumulation when applying iodine by means of foliar sprays. In contrast to soil applications, a higher iodine accumulation in the edible parts was observed using KI as an iodine fertilizer (Table 4.8) and the targeted iodine content was obtained at the lowest fertilizer rate of 0.5 kg I^--I ha^{-1}. Thus, the foliar fertilization technique was remarkably more efficient for the iodine biofortification of butterhead lettuce and it may be assumed that many leafy vegetables will show a similar response. Noteworthy, SMOLEŃ ET AL. (2011a) achieved a statistically significant difference in iodine accumulation spraying butterhead lettuce only at the highest fertilization dose of (4x) 2 kg IO_3^--I ha^{-1}. This may be explained by a rather high iodine level in the control treatment and due to surfactants being absent in the foliar spray solutions.

In the case of kohlrabi, satisfactory results in iodine accumulation could only be achieved by means of soil fertilization (Table 4.8). This may again be explained by the constitution and morphology of the two model systems: Kohlrabi has to translocate the adsorbed iodine from the leaf blades through the leaf stalks into the stem tuber, whereas butterhead lettuce can absorb iodine directly from the sprayed solution into the leaf cells. The low iodine levels found in the kohlrabi stem tuber indicate once more the marginal iodine phloem mobility. Kohlrabi and vegetable species with a similar morphology, i.e. with a clear separation between leaves and edible plant parts are not likely to be suitable for iodine foliar applications.

In conclusion, the very good iodine accumulation achieved in the case of butterhead lettuce and the similar expectations for leafy vegetables, as well as the lower raw material costs and the easy and inexpensive application of the foliar sprays, substantiated further trials with the focus on the foliar application in the experimental seasons 2011 - 2012.

5 Iodine biofortification by means of foliar sprays

5.1 Abstract

The iodine biofortification of vegetables by means of foliar sprays was investigated in field and greenhouse experiments. Aerial applications were ascertained to be a powerful and effective method of enhancing the iodine levels in several vegetable crops. With increasing iodine supply (0, 0.1, 0.25, 0.5, 0.75, 1.0 and 2.0 kg I ha^{-1}) to above-ground plant parts, the iodine concentrations in edible plant parts increased when vegetables were sprayed with KI and KIO_3 on different dates close to the harvest. The highest iodine accumulation was observed in leafy vegetables (winter spinach, butterhead lettuce, rocket, basil, parsley and oregano) and the desired iodine content of 50 - 100 µg I $(100\ g\ FM)^{-1}$ was already obtained at low fertilization rates of 0.1 - 0.25 kg I ha^{-1} without a significant yield reduction or degradation in the marketable quality.

Iodine applications under controlled environmental conditions in greenhouses demonstrated a comparable foliar uptake between KI and KIO_3 treatments. However, further field trials using butterhead lettuce as a model system showed greater efficacy of potassium iodide under certain circumstances. This could be explained by its higher hygroscopicity, the lower point of deliquescence and a smaller ionic size in comparison to potassium iodate. Different KI applications on butterhead lettuce during the course of the day showed the highest iodine concentration around midday and the wide iodine accumulation differences observed under fluctuating environmental conditions could be explained by the impact of light radiation, the relative humidity and the temperature conditions at the time of application. Furthermore, iodine treatments at different application dates nearing harvest showed an increasing iodine concentration in butterhead lettuce that could be associated to the rising fresh mass and to the increasing leaf area.

The leaf topography of some vegetable crops was examined and the stomatal size and distribution density was found to influence the absorption of aerially applied iodine solutions. Moreover, iodine was estimated to be marginally phloem mobile.

The miscibility of iodine with some agrochemicals was positively tested and tank mixing adjudged to work well with other agrochemicals. When KI or KIO_3 were sprayed simultaneously with commercial calcium fertilizers, fungicides or insecticides, iodine accumulation in butterhead lettuce was not affected and in some cases was even significantly enhanced. The latter result was probably due to an optimized foliar spray formulation of the applied products.

Additionally, the application of iodine foliar sprays at a dose relevant for practical implementation was found to slightly enhance the content of the total phenolic compounds and thus may improve the nutritional quality of multi-leaf lettuce as well.

5.2 Introduction

Studies on the wet deposition of radioiodine (^{125}I and ^{131}I) on vegetable leaves revealed the general iodine absorption ability of above-ground plant parts of different crop species (Oestling et al. 1989). Accordingly, Altinok et al. (2003) conducted successful iodine biofortification by means of foliar applications using potassium iodide on alfalfa plants. Smoleń et al. (2011a, 2011b) performed research on butter head lettuce and carrot applying potassium iodate as a foliar fertilizer. Both KI and KIO_3 foliar sprays were used by Strzetelsky et al. (2010) with partially satisfactory results on radish. However, in contrast to soil fertilization, currently no information is available on the response of different vegetable crops treated with both iodine forms applied at varied concentrations. In addition, only little is known about the appropriate application point in time for the foliar fertilization with iodine (i.e. during growing season, cultivation period and course of the day) as well as about the impact of exogenous factors (e.g. application equipment and climatic conditions) on the efficiency of this biofortification technique.

Under practical conditions in greenhouse and field production, foliar iodine treatments of horticultural crops may be realized in different ways: Firstly, by using overhead irrigation systems and thus combining the supply of water and fertilizers (= fertigation). Secondly, by means of mounted field sprayers using straight iodine sprays or the simultaneous application of iodine salts and other agrochemicals (= tank-mixing).

The water application rates, using mounted field sprayers, commonly range between 300 and 1000 L ha^{-1}; for a fertigation by means of overhead irrigation systems, they can vary widely, from several hundred up to 100,000 L ha^{-1}, depending on the specific needs of the grower, the climatic conditions, the cultivated crop and the respective cultivation system as well as the chosen irrigation intervals. However, a water application rate of 4,000 L ha^{-1} was deemed to be a reasonable value, especially in view of the conservation of water resources and the improved fertigation models suggested over the last years (VOOGT 2005; BECK 2009; HEMMING ET AL. 2009; KNECHT 2014).

In households, raw vegetables are commonly washed under flowing tap water before being further processed for human nutrition. Hence, the influence of this postharvest treatment on the iodine content in edible plant parts should be taken into account. Moreover, the comparison of washed to unwashed samples would allow for the determination to what extent the applied iodine was actually absorbed by the plant surface.

To improve operational processes, growers often mix different fertilizers or fertilizers and pesticides if the compatibility of the compounds allows blending (GRIFFITH 2010). Likewise, no information is available on the miscibility of iodine with different agrochemicals. Commercial iodine fertilizers for food crops are currently not available; however, some products for the use on pastures have been introduced in order to supply the grazing cattle with iodine and other essential trace elements (RAVENSDOWN 2013, YARA 2014b). Based on the implemented technology and experience in this area, iodine fertilizers for biofortification purposes might be developed with reasonable efforts.

Recent investigations indicate that iodine biofortification can significantly enhance the biosynthesis of important secondary metabolites such as flavonoids and anthocyanins in butter head lettuce or vitamin C in water spinach (BLASCO ET AL. 2008; WENG ET AL. 2008b). Lettuces, especially red varieties, are recognized as producing large quantities of anthocyanins and plant polyphenols which are known to have anti-oxidizing, anti-inflammatory, anti-diabetic and anti-cancer effects in the human body (LIU ET AL. 2007; PANDEY AND RIZVI 2009). Salanova® multi-leaf lettuce is, in turn, usually commercialized in a convenience packaging unit composed of green and red varieties which is becoming more and more popular, also because of its one-cut-ready feature (RIJK-ZWAAN 2009, 2012). The determination of total phenolic compounds in multi-leaf lettuce in the context with iodine biofortification studies would be an opportunity to better understand if iodine, applied at a reasonable dose for practical implementation, can improve this important quality trait.

Therefore, the main objective of this section was to conduct a series of foliar iodine fertilization trials under field and greenhouse conditions in order to:

- Compare a selection of vegetable species in their iodine accumulation behavior as affected by a single or twofold iodine foliar spray at different concentrations.
- Evaluate different culinary herb species in their iodine accumulation behavior when grown under controlled ambient conditions in the greenhouse and to investigate the iodine absorption into leaves by implementing different postharvest treatments.
- Determine an appropriate iodine dose for routine foliar fertilization and compare the efficiency and adequacy of the iodine forms I^- and IO_3^-.
- Assess the efficiency of an iodine fertilizer prototype versus pure grade KIO_3.
- Ascertain the miscibility of iodine with different agrochemicals and the effect of the application point in time and mode using butter head lettuce as a model system.
- Investigate the possible enhancement of total phenolic compounds in Salanova® green and red varieties as affected by a routine dose of iodine foliar sprays.

5.3 Experimental setup

The foliar application trials in season 2011 were conducted at different vegetable grower farms and greenhouse locations within a radius of approximately 100 km around Osnabrück, Germany. The main focus was on the iodine accumulation behavior of different field vegetables species (butterhead lettuce, crisp lettuce, winter spinach, wild rocket, broccoli and white cabbage) as affected by iodine foliar sprays at different concentrations (see trial setup in Table 5.1). Moreover, the performance of iodine foliar sprays at different concentrations (see trial setup in Table 5.2) was tested on some culinary herbs (parsley, chives, oregano and basil) under controlled ambient conditions (Figure 5.2) in the greenhouse at an application rate of 4,000 L H_2O ha^{-1}. In addition, the herb samples were subject to a postharvest treatment: each sample was separated into two homogeneous aliquots and then either washed under flowing tap water or not (Figure 5.6) before being further processed for iodine determination.

The trials conducted in 2012 on butterhead lettuce (cv. `Mafalda´) were intended to investigate the application of iodine fertilizers during the cultivation period and during the course of the day. In addition, the tank mixing of iodine (pure salts and fertilizer prototype) with agrochemicals (commercial calcium fertilizers and pesticides) was tested. Furthermore, the effect of iodine sprays on the phenolic compounds of Salanova® multi-leaf lettuce was investigated. All trials run in season 2012 were conducted on the arable fields of Mählmann Gemüsebau GmbH & Co. KG (Cappeln, Germany; N 52° 49' 21.657" E 8° 8' 29.678" and N 52° 49' 29.779" E 8° 9' 29.39"). An overview of the individual trial setup is given in Table 5.3. Briefly, the iodine sprays were applied once or twice as potassium iodide or potassium iodate solutions at various quantities (0.125 - 0.25 kg I ha^{-1}), water application rates (300 - 1,000 L H_2O) and points in time before harvest (1 - 14 days). The agrochemicals for the tank mixing trials, for example the insecticide Karate® and the fungicide Revus®, were provided by Mählmann Gemüsebau GmbH & Co. KG; the calcium fertilizers Calcinit® [$Ca(NO_3)_2$] and Stopit® ($CaCl_2$) as well as the iodine fertilizer prototype were provided by Yara GmbH & Co. KG., Dülmen, Germany.

Climatic field and greenhouse conditions and the respective cultivation periods of the crops in the trial seasons 2011/2012 are given in the Figures 5.1, 5.2 and 5.3. For an increased trial precision, the number of repetitions was elevated from n = 4 in 2011 to n = 5 in 2012. The gross plot area of all field trials was 4.5 m^2 (1.5 x 3.0 m) except for when spraying with a tractor mounted field sprayer (large scale application on 200 m^2 plots; cf. treatment VI in Table 5.3). In the greenhouse, the net plot area was 1 m^2. The plant material, conventional pesticides and fertilizers as well as the organic fertilizers and biological control agents were provided on-site by the respective growers. The leaf area (see Figure 5.11 B) was determined by using a digital camera (model DSC-W350, Sony Corporation, Tokyo, Japan) and an image analyzing program (SigmaScan® Pro, Systat Software Inc., San Jose, CA, USA). Details on statistical procedures, application techniques for iodine salts and sampling methods, as well as the analytical procedure for iodine determination are described in chapter 3.

5.3.1 Determination of total calcium in plant matrix

In order to detect possible differences in the calcium content of butterhead lettuce after spraying iodine salts mixed with calcium fertilizers (vide supra), total Ca in the plant matrix was determined following the dry ashing procedure described by CAMPBELL AND PLANK (1998). 2.00 g of milled and oven-dried plant material were weighed into 15 mL porcelain crucibles and heated on a hot plate until cessation of smoke emission and subsequently placed into a muffle furnace at 550 °C for 16 h. After cooling, 10 mL of 2 *M* HCl were added as an ashing aid and the crucibles were heated again on a hot plate until boiling. The cooled and solubilized ash was then quantitatively transferred in 100 mL volumetric flasks by rinsing the crucibles with deionized water (the solution was filtered through folded filters type MN 614 G ¼). Ca-detection occurred by means of AES running on the acetylene gas mode (2400 °C) as described for the K-detection in section 3.4.3.

Table 5.1 Overview of trial setup and growing conditions of field crops in season 2011. Experiment no. 5: Comparison of the iodine accumulation behavior of different vegetable crops as affected by iodine foliar sprays at varied concentrations

Exp. no.	Grower	Arable field location	Vegetable species	Treatments	Growing conditions: Basic fertilization N [kg ha^{-1}]	P_2O_5 [kg ha^{-1}]	K_2O [kg ha^{-1}]	Weed management and pest control
5	**Gemüsehof Biewener KG**, Telgheide 28, 49328 Melle, Germany	N 52° 15' 18.087" E 8° 25' 42.401"	Butterhead lettuce cv. \`Mafalda´ (I = early planting)	Control 0 kg I ha^{-1}, KI at 0.5, 0.75, 1 kg I ha^{-1}, KIO_3 at 0.5, 0.75, 1 kg I ha^{-1}	120	40	160	Kerb® FLO 3.75 L ha^{-1}, Signum® 1.5 kg ha^{-1}, Stomp Aqua ® 1.8 L ha^{-1}, Teldor® 1.0 kg ha^{-1}
		N 52° 13' 58.057" E 8° 26' 46.208"	Butterhead lettuce cv. \`Mafalda´ (II = midseason plant.)		No mineral fertilization occurred (green manure and crop residues)			Kerb® FLO 3.75 L ha^{-1}, Signum® 1.5 kg ha^{-1} Stomp Aqua® 1.8 L ha^{-1}, Teldor® 1.0 kg ha^{-1}
		N 52° 14' 56.333" E 8° 24' 48.637"	Butterhead lettuce cv. \`Mafalda´ (III = late planting)	Control 0 kg I ha^{-1}, KIO_3 as pure salt or fertilizer prototype at 0.5, 0.75, 1 kg I ha^{-1}	No mineral fertilization occurred (large crop residues → EHEC crisis)			Kerb® FLO 3.75 L ha^{-1}, Calypso® 0.2 L ha^{-1}, Karate® Zeon 0.07 L ha^{-1}, Acrobat® 2 kg ha^{-1}, Ortiva® 1 L ha^{-1}, Steward® 0.08 L ha^{-1}, Signum® 1.5 kg ha^{-1}
	Mählmann Gemüsebau GmbH & CO. KG, Im Siehenfelde 13, 49692 Cappeln, Germany	N52° 44' 57.135" E8° 16' 2.426"	Crisp lettuce cv. \`Argentinas´	Control 0 kg I ha^{-1}, KI at 0.5, 0.75, 1 kg I ha^{-1}, KIO_3 at 0.5, 0.75, 1 kg I ha^{-1}	141	0.7	209	Ridomil® Gold SZ 2 kg ha^{-1}, Signum® 1.5 kg ha^{-1} Fastac® SC 0.09 L ha^{-1}, Plenum® WG 400 g ha^{-1} Acrobat® 2 kg ha^{-1}, Karate® Zeon 0.07 L ha^{-1}
		N 52° 49' 11.481" E 8° 14' 14.28"	Winter spinach cv. \`Falcon´		199	0	60	Goltix® 700 SC 0.8 L ha^{-1}, Betosip® SC 1.0 L ha^{-1} Asket® 470 0.4 L ha^{-1}
		N 52° 50' 16.543" E 8° 7' 35.064"	Broccoli (I; II) cv. \`Marathon´	Control 0 kg I ha^{-1}, KI at 0.5, 0.75, 1 kg I ha^{-1}, KIO_3 at 0.5, 0.75, 1 kg I ha^{-1}, (sprayed I = 1x or II = 2x)	200	0	160	Butisan® 1.5 L ha^{-1}, Pirimor® 250 g ha^{-1} Equation® 500 g ha^{-1}, Decis® 0.2 L ha^{-1} Forum® 1.2 L ha^{-1}, Folicur® 0.5 L ha^{-1} Bulldock® 0.3 L ha^{-1}, Karate® Zeon 0.075 L ha^{-1}
	Gemüsehof Andreas Wehmeyer, Rödgerei 8, 32051 Herford, Germany	N 52° 5' 16.958" E 8° 37' 15.351"	White cabbage cv. \`Impala´	Control 0 kg I ha^{-1}, KI at 0.5, 0.75, 1 kg I ha^{-1}, KIO_3 at 0.5, 0.75, 1 kg I ha^{-1}, (sprayed 2x)	246	0	240	Butisan® 1.5 L ha^{-1}, Signum® 1.5 kg ha^{-1} Folicur® 1 L ha^{-1}, Calypso® 0.2 L ha^{-1} Karate® Zeon 0.075 L ha^{-1}, Fastac® 0.09 L ha^{-1} Perfekthion® 0.6 L ha^{-1}, SpinTor® 0.2 L ha^{-1} Steward® 0.085 L ha^{-1}, Plenum® WG 400 g ha^{-1}
	Stefan Stegemeier Gemüsebau, Blackenfeld 148, 33739 Bielefeld, Germany	N 52° 4' 20.204" E 8° 34' 31.393"	Wild rocket cv. \`Tricia´	Control 0 kg I ha^{-1}, KI at 0.5, 0.75, 1 kg I ha^{-1}, KIO_3 at 0.5, 0.75, 1 kg I ha^{-1}	180	0	120	Thermal weed extermination by water vapor device, Karate® Zeon 0.07 L ha^{-1}, Acrobat® 2 kg ha^{-1}

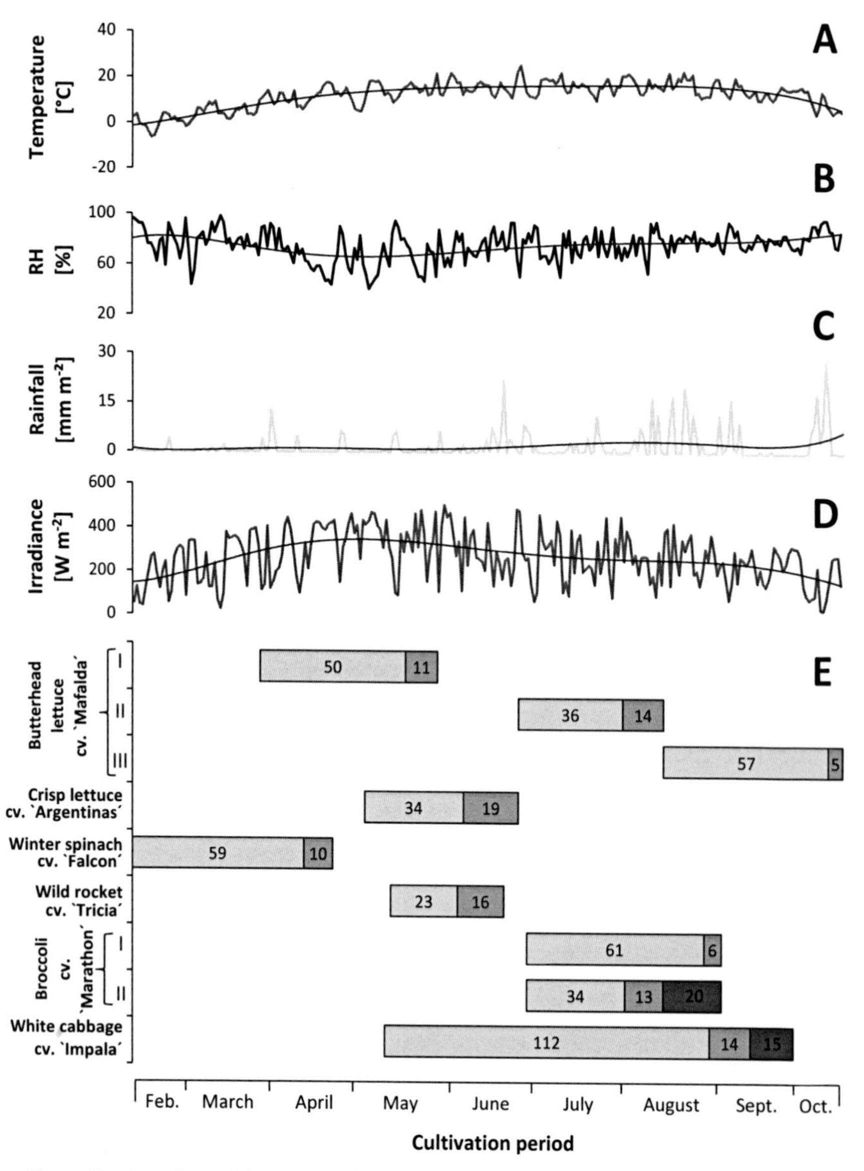

Figure 5.1 Overview of climatic conditions during cultivation period and respective growth period of different vegetables. Climatic data collected at Schwaney meteorological station, Germany, and calculated as daily averages. A = air temperature; B = relative humidity; C = rainfall; D = irradiance; E = relative cultivation time in days and cultivation time-slot of vegetables investigated in season 2011 (2011.02.11 - 2011.10.17). Light grey bars indicate days from planting/sowing to 1st foliar application; middle grey bars indicate days from 1st application to harvest or 2nd application; dark grey bars indicate days from 2nd application to harvest

Table 5.2 Overview of trial setup and growing conditions of potted herbs grown in greenhouses in season 2011. Experiment no. 6: Comparison of the iodine accumulation behavior of different herbs species grown in greenhouses as affected by iodine foliar sprays at varied concentrations. Herbs grown in containers (13 cm ø) filled with peat substrate (Klaasmann-Deilmann GmbH, Germany; KKS Bio-Kräutersubstrat: pH 5.8, total organic nitrogen = 700 mg N L^{-1}, 300 - 400 mg P_2O_5 L^{-1}, 400 - 650 mg K_2O L^{-1} and 100 - 150 mg Mg L^{-1}). Municipal water (EC = 0.2, pH = 7.8, total hardness = 0.65 mmol L^{-1}, free CO_2 = 2.4 mg L^{-1}, NH_4 = 0.05 mg L^{-1}, K = 2.4 mg L^{-1}, Ca = 22 mg L^{-1}, Mg = 2.4 mg L^{-1}) and OPF fertilizer (Organic plant feed, Plant Health Care BV, The Netherlands; total N = 82.2 g L^{-1}, P_2O_5 = 30.3 g L^{-1}, K_2O = 32.1 g L^{-1}, S = 0.4 g L^{-1}, B = 7 mg L^{-1}, Ca = 390 mg L^{-1}, Cu = 2 mg L^{-1}, Fe = 120 mg L^{-1}, Mg = 850 mg L^{-1}, Mn = 15 mg L^{-1}, Zn = 12 mg L^{-1}) were used for fertigation. Fertigation frequency and time were adjusted depending on water consumption

Exp. no.	Grower	Green-house location	Herb species	Treatments	Growing conditions			
					Fertigation			Pest control
					EC [mS cm^{-1}]	Stock A	Stock B	
6	**Friedrich Schultz Gartenbau GmbH & Co. KG,** Großes Meer 6, 26871 Papenburg, Germany	N 53° 1' 24.841" E 7° 24' 11.163"	Parsley cv. \`Rina´ (*Petroselinum crispum* [Mill.] Nym.)	Control 0 kg I ha^{-1}, KI at 0.1, 0.2 and 0.5 kg I ha^{-1}, KIO_3 at 0.1, 0.2 and 0.5 kg I ha^{-1} (sprayed at a rate of 4000 L H_2O ha^{-1})	1.5 - 1.7	250 mL OPF L^{-1}, 37.5 g sulfate of potash magnesia L^{-1} (30 % K_2O, 10 % MgO, 18 % S)	12.5 g epsomite L^{-1} (16 % Mg, 13 % S), 1.5 mL Lebosol® B L^{-1}, 0.225 mL Lebosol® Zn L^{-1}, 1.5 mL Lebosol® Mn L^{-1}, 0.075 mL Lebosol® Cu L^{-1}, 2 g Bolikel® Fe L^{-1}	*Bacillus thuringiensis* subsp. *israelensis* (BioMükk® WDG, Biofa AG, Germany)
			Chives cv.\`Dolores´ (*Allium schoenoprasum* L.)					
	Gartenbau Borrmann, Deverhof 11, 26871 Papenburg, Germany	N 53° 2' 6.49" E 7° 23' 20.759"	Origano cv. \`Dora´ (*Origanum vulgare* L.)		1.5 - 1.7	500 mL OPF L^{-1}, 75 g sulfate of potash magnesia L^{-1} (30 % K_2O, 10 % MgO, 18 % S), 12.5 g epsomite L^{-1} (16 % MgO, 13 % S), 2 mL Lebosol® B L^{-1}, 0.5 mL Lebosol® Zn L^{-1}, 2 mL Lebosol® Mn L^{-1}, 0.15 mL Lebosol® Cu L^{-1}, 5 g Tenso® Fe L^{-1}		*Steinernema feltiae* (Biofa AG, Germany), *Aphidius colemanii*, *Chrysoperla carnea* (Sautter und Stepper GmbH, Germany), *Amblyseius cucumeris* (Katz Biotech AG, Germany)
			Basil cv. \`Hannah´ (*Ocimum basilicum* L.)		2.2 - 2.5			

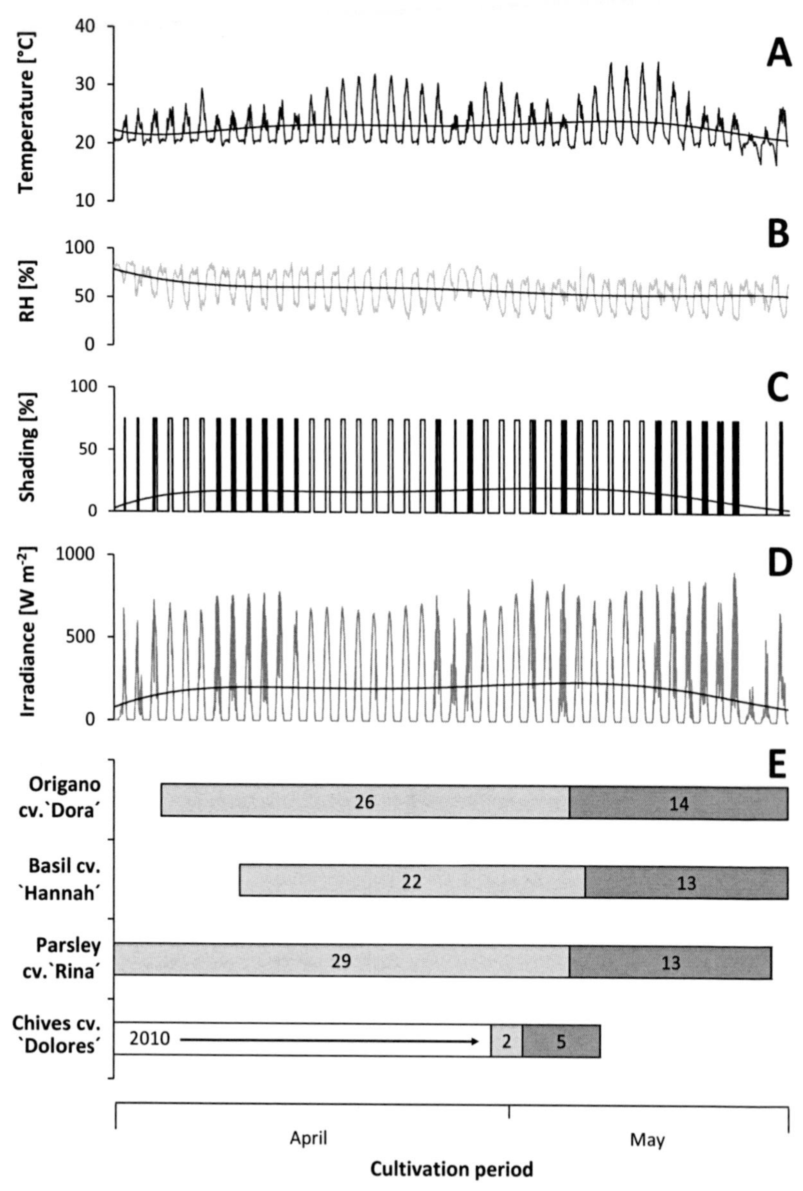

Figure 5.2 Overview of greenhouse conditions during cultivation period and respective growth period of different potted herbs. Data collected in Papenburg, Germany at a rate of 144 measuring points per day (average of two greenhouses). A = air temperature; B = relative humidity; C = shading; D = irradiance; E = relative cultivation time in days and cultivation time-slot of potted herbs (2011.04.06 - 2011.05.18). Light grey bars indicate days from propagation to foliar application; dark grey bars indicate days from foliar application to harvest; white bar indicates freezing until propagation (preparatory culture in season 2010)

Table 5.3 Overview of trial setup and growing conditions in season 2012. Experiment no. 7: Miscibility of agrochemicals with iodine foliar sprays and influence of exogenous factors on application. Experiment no. 8: Influence of iodine foliar sprays on iodine accumulation and total phenolic compounds of Salanova® multi-leaf lettuce. Application rate of foliar sprays = 1000 L H_2O ha^{-1} (with exception of treatment V)

Exp. no.	Vegetable species	Treatments			Growing conditions: Basic fertilization N [kg ha^{-1}]	P_2O_5 [kg ha^{-1}]	K_2O [kg ha^{-1}]	Weed management and pest control
7	Butterhead lettuce cv. `Mafalda´	I	0 kg I ha^{-1}, KIO_3 at 0.125 kg I ha^{-1} as pure salt or fertilizer prototype (EC)	Sprayed 2x in combination with Ca-fertilizers: Cacinit® or Stop-it ® at 1 kg Ca ha^{-1}	120	0	160	Cultivation guard nets; Kerb® FLO 3.75 L ha^{-1}, Signum® 1.5 kg ha^{-1}, Acrobat® 2 kg ha^{-1}, Fastac® SC 0.09 L ha^{-1}
		II	0 kg I ha^{-1} 0.25 kg I^--I ha^{-1} 0.25 kg IO_3^--I ha^{-1}	Sprayed in combination with pesticides: Revus® at 600 mL ha^{-1} or Karate® Zeon at 75 mL ha^{-1}	60	0	80	Revus® 600 mL ha^{-1}, Karate® Zeon 75 mL ha^{-1}, Kerb® FLO 3.75 L ha^{-1}, Signum® 1.5 kg ha^{-1}
		III	0 kg I ha^{-1} 0.25 kg I^--I ha^{-1} 0.25 kg IO_3^--I ha^{-1}	Sprayed at 7:00, 11:00, 15:00 and 19:00 hours (Control only once at 11:00 hours)	60	0	80	Cultivation guard nets; Kerb® FLO 3.75 L ha^{-1}, Signum® 1.5 kg ha^{-1}, Acrobat® 2 kg ha^{-1}, Plenum® WG 400 g ha^{-1}
		IV	0 kg I ha^{-1} 0.25 kg I^--I ha^{-1} 0.25 kg IO_3^--I ha^{-1}	Sprayed 14, 7 and 1 days before harvest (Control only once, 1 day before harvest)	120	0	160	Kerb® FLO 3.75 L ha^{-1}, Ortiva® 1 L ha^{-1}, Karate® Zeon 0.075 L ha^{-1}, Acrobat® 2 kg ha^{-1}, Calypso® 0.2 L ha-1, Steward ® 0.08 L ha^{-1}
		V	0 kg I ha^{-1} 0.25 kg I^--I ha^{-1} 0.25 kg IO_3^--I ha^{-1}	Sprayed at 300, 600 and 1000 L H_2O ha^{-1} (Control only once at 600 L H_2O ha^{-1})	120	0	160	Kerb® FLO 3.75 L ha^{-1}, Ortiva® 1 L ha^{-1}, Karate® Zeon 0.07 L ha^{-1}, Acrobat® 2 kg ha^{-1}, Calypso® 0.2 L ha^{-1}
		VI	0 kg I ha^{-1} 0.25 kg I^--I ha^{-1} 0.25 kg IO_3^--I ha^{-1}	Sprayed by means of mounted field sprayer (Amazone UF 1201) or by hand-held spray system (Easy Sprayer Plus; Lehnartz GmbH) (Control only by hand-held spray system)	60	0	80	Cultivation guard nets; Kerb® FLO 3.75 L ha^{-1}, Signum® 1.5 kg ha^{-1}, Acrobat® 2 kg ha^{-1}, Fastac® SC 0.09 L ha^{-1}
8	Salanova® multi-leaf lettuce cv. `Archimedes RZ´ (green variety) and cv. `Gaugin RZ´ (red variety)	VII	0 kg I ha^{-1} 0.25 kg I^--I ha^{-1} 0.25 kg IO_3^--I ha^{-1}		120	0	160	Signum® 1.5 kg ha^{-1}, Acrobat® 2 kg ha^{-1}, Kerb® FLO 3.75 L ha^{-1}, Revus® 600 ml ha^{-1}, Fastac® SC 0.09 L ha^{-1}, Plenum® WG 400 g ha^{-1}, Karate® Zeon 75 mL ha^{-1}

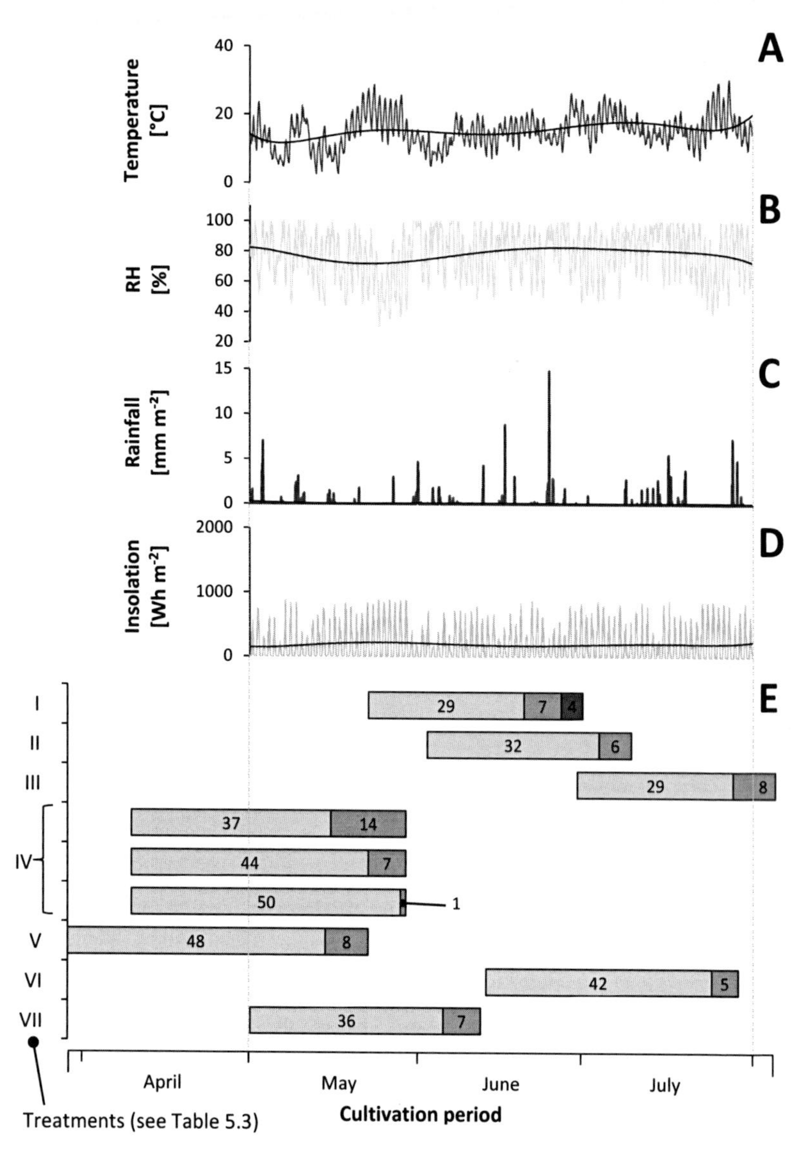

Figure 5.3 Overview of climatic conditions (2010.05.01 - 2010.07.31) during foliar application and respective growth period. Climatic data collected at the meteorological station of Emstek, Germany, at a rate of 24 measuring points per day. A = air temperature; B = relative humidity; C = rainfall; D = insolation; E = relative cultivation time in days and cultivation time-slot of vegetables investigated in season 2012 (2012.03.28 - 2012.08.04). Light grey bars indicate days from planting to 1st foliar application; middle grey bars indicate days from 1st application to harvest or 2nd application; dark grey bars indicate days from 2nd application to harvest

5.3.2 Determination of total phenolic compounds

In view to detect a possible influence of iodine sprays on the production of secondary plant compounds, the total phenolic content in multi-leaf lettuce was determined (cf. 5.2). The first preparative step was the freeze drying of the plant matter samples in a lyophilization system (model P22K-E-6, Dieter Piatkowski - Forschungsgeräte, München, Germany) following the DIN EN ISO 16720 method (2007b). The plant material was then finely ground by using a 250 µm sieve in an ultra-centrifugal rotor mill (cf. 3.3.1) and stored in 250 mL plastic boxes in a desiccator at room temperature with no direct sunlight influence until extraction.

The extraction was conducted as follows: 0.5000 g of the prepared plant material was filled in 15 mL PP-tubes (Sarstedt AG & Co., Nümbrecht, Germany) and 10 mL of cooled ethanol (80 %) was added. The contents of the closed tubes were then mixed on a Vortex-shaker for 1 minute and placed in an ultrasonic bath for a further 20 minutes. The samples were then centrifuged at 5,000 g for 10 minutes in a centrifuge model 5804 R (Eppendorf AG, Hamburg, Germany). The supernatants were decanted separately in a graduated flask and the sample residues were dissolved in ethanol again, in the same manner as described above. The two acquired supernatants (from each sample) were then combined in the graduated flask and filled up to 20 mL with ethanol. The samples were extracted in a threefold replication. The determination of the total phenolics was performed on a spectral photometer model HP 8453 (Hewlett-Packard Company, Palo Alto, CA, USA) following the microscale protocol for the Folin-Ciocalteu colorimetry (WATERHOUSE 2002), which is based on the key methods suggested by SINGLETON AND ROSSI (1965), SOMERS AND ZIEMELIS (1985) and SINGLETON ET AL. (1999). The main modifications to this method were: The measuring of the sample absorbance at 725 nm instead of 765 nm and the use of quercitin instead of gallic acid as a calibration standard (0, 50, 100, 250 and 500 mg L^{-1}); the results were expressed as milligram quercitin equivalents per gram of dry matter [mg QE $(g\ DM)^{-1}$].

5.3.3 Microscopic methods

Specific morphological and structural traits of plants (e. g. stomata or trichomes) may significantly influence the uptake of nutrient solutions by the leaf surface (FERNÁNDEZ ET AL. 2013). Hence, preliminary microscopic investigations were conducted in order to better understand the impact of the plant leaf topography on the iodine uptake of some vegetable crops.

The specimens for the examinations by means of light microscopy (LM) were prepared according to the technique described by KREEB (1990). Transparent nail varnish was applied drop-wise to the leaf samples and allowed to dry. The epidermal impressions were then peeled off with a scalpel, mounted on a slide and, after adding a drop of deionized water, covered with a coverslip. The examination was performed with a transmitted light microscope (model BX 61, Olympus Corporation, Tokyo, Japan) equipped with a camera (model ColorView III, Soft Imaging System GmbH, Münster, Germany). The size of the field of view was calibrated using a stage micrometer and the number of stomata per square millimeter assigned to classes (< 50, 50 - 75, 75 - 100 and > 100 stomata mm^{-2}).

The specimens for the examinations by means of scanning electron microscopy were prepared using a cryogenic system (model K 1250 X, Quorum Technologies Ltd., Lewes, UK). Rectangular leaf sections of approx. 5 x 10 mm in size were cut out at random from the vegetable leaves and affixed to the multipurpose stub of the transfer rod. The samples were transferred to the cryo-unit and placed in a vacuum, then deep-frozen at -150 °C using liquid nitrogen and subsequently sublimated at -80 °C to achieve a freeze-dried examination surface. The leaf surfaces were then sputtered with a carbon layer (30 sec. at 20 mA) and transferred with the transfer rod to a scanning electron microscope (model Zeiss Auriga®, Carl Zeiss AG, Oberkochen, Germany) equipped with a low temperature device.

5.4 Results

5.4.1 Experiment no. 5: Comparison of the iodine accumulation behavior between different vegetable crops as affected by iodine foliar sprays at varied doses and forms

The influence of single or twofold iodine foliar sprays (0 - 1.0 kg I ha^{-1}) on different vegetable crops was investigated using KI and KIO_3 as fertilizer. Butterhead lettuce, rocket, broccoli and white cabbage were found to tolerate all aerially applied doses in both iodine forms well and no statistically significant differences in biomass production were found. Severe leaf burn was observed on crisp lettuce and winter spinach (Figure 5.4 and 5.5) although biomass production was affected significantly only in the case of crisp lettuce (Table 5.4). Remarkably, IO_3^- treatments on crisp lettuce were observed to induce more leaf impairment than I^- treatments (Figure 5.4 E and F compared to C and D), especially at the highest dose of 1 kg I ha^{-1}. Furthermore, leaf burn was found only on the outer leaves of crisp lettuce. This part of the foliage usually remains as a crop residue on the arable field. Concordantly, the harvested produce was not impaired in its marketable quality. On the leaves of winter spinach, a similar leaf burn was noticed when comparing the two iodine forms (Figure 5.5 B compared to C and D compared to E) and a clear severity gradient in leaf burn was observed with increasing fertilizer concentration (Figure 5.5 B and C compared to D and E).

Table 5.4 Yield of different vegetable crops as affected by iodine foliar sprays at varied concentrations. Yield expressed as percent of unfortified control treatment. Percentages with same letters do not differ according to the Bonferroni MCP at $\alpha = 0.05$. Levels of significance are represented by * = $p < 0.05$, ** = $p < 0.01$, *** = $p < 0.001$ and NS = not significant = $p > 0.05$ (actual probability level). + = two factorial GLM ANOVA: Block = ** (0.001206), iodine form = ** (0.001347), iodine dose = NS (0.418367), F x D = NS (0.291534). † = sprayed twice at the denoted treatment dose. ‡ = not normally distributed dataset; parameter free test methods were used instead of the Bonferroni MCP. IFP = iodine fertilizer prototype. n = 4

Cultivated plant species	Butterhead lettuce I (Early planting)		Butterhead lettuce II (Midseason planting)		Butterhead lettuce III (Late planting)		Winter spinach		Rocket	
Treatment dose [kg I ha^{-1}] / Iodine form [I^- / IO_3^-]	I^-	IO_3^-	I^-	IO_3^-	IO_3^- (IFP)	IO_3^-	I^-	IO_3^-	I^-	IO_3^-
	Relative crop yield [%]									
0	100.0 a		100.0 a		100.0 a		100.0 a		100.0 a	
0.5	92.5 a	104.1 a	97.4 a	100.9 a	104.6 a	105.7 a	103.4 a	100.5 a	100.5 a	100.0 a
0.75	96.2 a	106.5 a	97.1 a	94.5 a	106.8 a	98.4 a	99.2 a	101.6 a	101.5 a	100.4 a
1.0	107.6 a	100.2 a	99.6 a	94.6 a	104.3 a	107.3 a	98.9 a	96.8 a	102.9 a	99.8 a
ONE WAY ANOVA (p-Value)	NS (0.781269)		NS (0.893943)		NS (0.650206)		NS (0.722853)		NS (0.902850)	

Cultivated plant species	Crisp lettuce (+)		White cabbage (†;‡)		Broccoli I Vegetative phase		Broccoli II Generative phase (†)	
Treatment dose [kg I ha^{-1}] / Iodine form [I^- / IO_3^-]	I^-	IO_3^-	I^-	IO_3^-	I^-	IO_3^-	I^-	IO_3^-
	Relative crop yield [%]							
0	100.0 a		100.0		100.0 a		100.0 a	
0.5	88.8 ab	83.0 ab	98.4 NS	96.7 NS	97.5 a	100.3 a	92.2 a	96.2 a
0.75	90.7 ab	77.2 b	96.1 NS	99.2 NS	97.5 a	91.9 a	100.8 a	90.8 a
1.0	85.4 ab	79.2 ab	96.0 NS	96.9 NS	98.5 a	94.9 a	100.2 a	91.9 a
ONE WAY ANOVA (p-Value)	* (0.032694)		Non-parametric tests: Wilcoxon rank sum test (compared to control) ↑ Friedman's test = NS (0.71)		NS (0.774015)		NS (0.578187)	

Figure 5.4 Visual comparison of crisp lettuce two weeks after foliar spray application at different iodine doses and forms: A = single leaf at 0 kg I^--I ha^{-1}, B = head at 0 kg I^--I, C = single leaf at 1 kg I^--I ha^{-1}, D = head (upside down) at 1 kg I^--I ha^{-1}, E = single leaf at 1 kg IO_3^--I ha^{-1}, F = head (upside down) at 1 kg IO_3^--I ha^{-1}

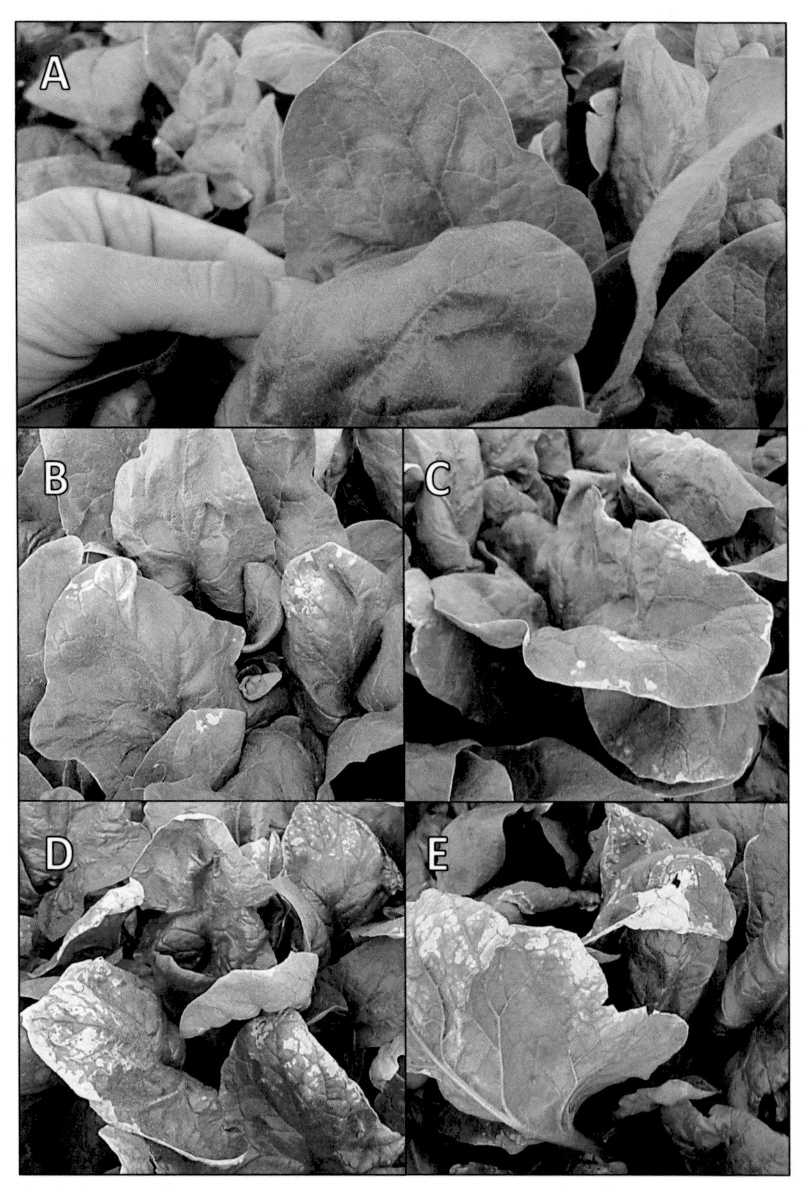

Figure 5.5 Visual comparison of winter spinach ten days after foliar spray application at different iodine doses and forms: A = 0 kg I ha^{-1}, B = 0.5 kg I^--I, C = 0.5 kg IO_3^--I ha^{-1}, D = 1 kg I^--I ha^{-1}, E = 1 kg IO_3^--I ha^{-1}

Table 5.5 shows the iodine accumulation behavior of different vegetable species after a single or twofold iodine application by means of foliar sprays. In butterhead lettuce, winter spinach and rocket, a distinct increase of the iodine concentration in edible plant parts was observed while raising the iodine dosage. On the contrary, in crisp lettuce, white cabbage and broccoli, only low iodine concentrations were detected in the edible plant parts, whereby crisp lettuce and broccoli showed no statistically significant differences. Remarkably, high iodine concentrations were found in the outer leaves of the crisp lettuce, whereas the outer leaves of the white cabbage accumulated minimally more iodine compared to the unfortified control. Furthermore, broccoli did not accumulate significant amounts of iodine even following a twofold foliar spray sprayed directly on the edible plant parts [generative phase (II)].

The successive planting of butterhead lettuce [early (I), midseason (II) and late (III) planting] showed very wide differences in iodine accumulation. The iodine concentration in the edible parts of summer crops cultivated in July/August (midseason planting) were up to 10 fold lower than in the early and late crops. Comparing the fertilizer prototype to the pure KIO_3 salt in the late cultivated lettuce, a slightly higher iodine accumulation was observed using the commercial formulation.

It is noteworthy, that in contrast to the soil fertilization trials, no clear superior iodine form could be determined. Fluctuations in the effectiveness of the iodine forms were observed throughout all foliar application trials. Nevertheless, the factors iodine form and dose were calculated to be significantly different in the ANOVA of all trials with the exception of broccoli. The factor plant part and growing phase were found throughout to be significantly different.

Table 5.5 The iodine content of different vegetable crops as affected by iodine foliar sprays at varied concentrations. Means expressed as µg I (100 g FM)$^{-1}$. Means with same letters do not differ according to Bonferroni MCP at $\alpha = 0.05$. Levels of significance are represented by * = $p < 0.05$, ** = $p < 0.01$, *** = $p < 0.001$ and NS = not significant = $p > 0.05$ (actual probability level). † = sprayed twice at the denoted treatment dose. ‡ = not normally distributed dataset; parameter free test methods were used instead of the Bonf. MCP. IFP = iodine fertilizer prototype. n = 4

Cultivated plant species 〉	Butterhead lettuce I (Early planting)		Butterhead lettuce II (Midseason planting)		Butterhead lettuce III (Late planting)		Winter spinach (‡)		Rocket	
Treatment dose [kg I ha^{-1}] / Iodine form [I^- / IO_3^-] 〉	I^-	IO_3^-	I^-	IO_3^-	IO_3^{-} (IFP)	IO_3^-	I^-	IO_3^-	I^-	IO_3^-
	Iodine concentration [µg I (100 g FM)$^{-1}$]									
0	0.4 a		4.5 a		9.3 a		11.7		5.5 a	
0.5	187.1 b	257.6 bc	51.3 bcd	31.2 ab	307.9 bc	222.4 b	494.3*	705.6*	62.6 bc	49.5 b
0.75	291.4 bc	384.0 cd	73.2 cd	38.9 abc	463.1 cd	259.2 b	782.4*	792.2*	87.1 bcd	64.3 bc
1.0	423.0 d	494.5 d	80.0 d	52.4 bcd	492.4 d	446.0 cd	1039.7*	1178.8*	125.9 d	99.8 cd
ONE WAY ANOVA	*** (0.000000)		*** (0.000000)		*** (0.000000)		Non-parametric tests: Wilcoxon rank sum test (compared to control) ↑; Friedman's test = * (0.039)		*** (0.000000)	
GLM ANOVA two factorial Block	NS (0.939719)		NS (0.071042)		NS (0.386898)				NS (0.152094)	
Iodine form (F)	*** (0.008871)		*** (0.000244)		** (0.002631)				** (0.008867)	
Iodine dose (D)	*** (0.000005)		* (0.015157)		*** (0.000529)				*** (0.000025)	
F x D	NS (0.875239)		NS (0.823213)		NS (0.105769)				NS (0.732189)	

Cultivated plant species 〉	Crisp lettuce				White cabbage (†)				Broccoli I Vegetative phase		Broccoli II Generative phase (†)	
Analyzed plant part >	Edible		Outer leaves		Edible		Outer leaves		Edible			
Treatment dose [kg I ha^{-1}] / Iodine form [I^- / IO_3^-] 〉	I^-	IO_3^-	I^-	IO_3^-	I^-	IO_3^-	I^-	IO_3^-	I^-	IO_3^-	I^-	IO_3^-
	Iodine concentration [µg I (100 g FM)$^{-1}$]											
0	1.4 a		10.3 a		1.5 a		0.2 a		3.5 a		4.5 a	
0.5	5.9 a	4.9 a	146.7 b	170.1 b	7.7 abc	4.7 ab	9.5 b	10.4 b	1.7 a	2.9 a	7.6 a	7.0 a
0.75	1.8 a	1.1 a	239.8 b	179.4 b	11.4 c	5.8 abc	12.5 b	12.9 b	2.9 a	4.1 a	5.4 a	7.5 a
1.0	10.3 a	1.7 a	462.4 c	219.3 b	11.6 c	8.1 bc	19.7 b	12.9 b	4.2 a	3.9 a	5.8 a	8.6 a
ONE WAY ANOVA	NS (0.058925)		*** (0.000000)		*** (0.000245)		*** (0.000000)		NS (0.471555)		NS (0.071172)	
GLM ANOVA three factorial Block	*** (0.000002)				NS (0.913066)				NS (0.239673)			
Iodine form (F)	*** (0.000000)				* (0.018671)				NS (0.054656)			
Iodine dose (D)	*** (0.000000)				* (0.011906)				NS (0.430040)			
Plant part / growing phase	*** (0.000000)				*** (0.00 0172)				*** (0.000000)			
F x D	*** (0.000000)				NS (0.711197)				NS (0.935427)			
F x P	*** (0.000762)				NS (0.108071)				NS (0.831719)			
D x P	*** (0.000000)				NS (0.910865)				NS (0.260698)			
F x D x P	*** (0.000149)				NS (0.403107)				NS (0.126222)			

5.4.2 Experiment no. 6: Comparison of the iodine absorption behavior of some potted herb species grown in greenhouses as affected by varied iodine foliar sprays

The influence of iodine foliar sprays on four different culinary herbs grown in greenhouse was investigated in experiment no. 6 using single sprays of 0 - 0.5 kg I ha^{-1} applied as KI and KIO_3. All potted herbs were found to tolerate the aerially applied iodine well. The marketable quality was not affected in any of the trials and the biomass production showed no significant differences (Table 5.6).

Table 5.6 The yield of different culinary herbs as affected by iodine foliar sprays at varied concentrations. Yield expressed as percent of unfortified control treatment. Percentages with same letters do not differ according to Bonferroni MCP at α = 0.05. Levels of significance are represented by * = p < 0.05, ** = p < 0.01, *** = p < 0.001 and NS = not significant = p > 0.05 (actual probability level). n = 4

Herb species	Oregano		Basil	
Treatment dose [kg I ha^{-1}] / Iodine form [I^- / IO_3^-]	I^-	IO_3^-	I^-	IO_3^-
	Relative crop yield [%]			
0	100.0 a		100.0 a	
0.1	95.8 a	96.5 a	97.2 a	103.3 a
0.2	97.2 a	96.6 a	99.7 a	99.0 a
0.5	96.2 a	92.0 a	100.5 a	99.9 a
ONE WAY ANOVA (p-Value)	NS (0.825117)		NS (0.476410)	
Herb species	**Parsley**		**Chives**	
Treatment dose [kg I ha^{-1}] / Iodine form [I^- / IO_3^-]	I^-	IO_3^-	I^-	IO_3^-
	Relative crop yield [%]			
0	100.0 a		100.0 a	
0.1	99.3 a	96.2 a	95.4 a	100.9 a
0.2	101.3 a	100.1 a	99.2 a	99.4 a
0.5	102.8 a	96.8 a	105.3 a	97.2 a
ONE WAY ANOVA (p-Value)	NS (0.215681)		NS (0.714872)	

Significant differences in the iodine accumulation behavior of potted herbs were observed (Table 5.7). Enhancing iodine contents in edible plant parts were found throughout with increasing iodine dose. All the potted herbs showed fairly high accumulation levels with a slightly higher variation within the iodide treatments. Once again, the results did not clearly define a better suited iodine form. Remarkably, the factor iodine form was found to be not significantly different in the cases of parsley and chives (GLM ANOVA in Table 5.7) whereas the factor iodine dose was significant throughout. The accumulation levels in basil and parsley were comparable whereas oregano showed the highest and chives the lowest iodine uptakes.

Table 5.7 Iodine concentration of different potted herbs as affected by iodine foliar sprays at varied concentrations. Means expressed as µg I (100 g FM)$^{-1}$. Means with same letters do not differ according to Bonferroni MCP at $\alpha = 0.05$. Levels of significance are represented by * = $p < 0.05$, ** = $p < 0.01$, *** = $p < 0.001$ and NS = not significant = $p > 0.05$ (actual probability level). n = 4

Herb species		**Oregano**		**Basil**	
Treatment dose [kg I ha^{-1}]	Iodine form [I^- / IO_3^-]	I^-	IO_3^-	I^-	IO_3^-
		Iodine concentration [µg I (100 g FM)$^{-1}$]			
0		4.2 a		9.2 a	
0.1		139.2 b	113.4 b	63.9 b	91.9 b
0.2		172.3 b	175.2 b	118.4 bc	185.9 c
0.5		613.0 d	371.1 c	353.0 d	443.8 d
ONE WAY ANOVA		*** (0.000000)		*** (0.000000)	
GLM ANOVA two factorial	Block	NS (0.344922)		NS (0.502348)	
	Iodine form	** (0.001121)		*** (0.000612)	
	Iodine dose	*** (0.000000)		*** (0.000000)	
	F x D	** (0.002285)		NS (0.658367)	
Herb species		**Parsley**		**Chives**	
Treatment dose [kg I ha^{-1}]	Iodine form [I^- / IO_3^-]	I^-	IO_3^-	I^-	IO_3^-
		Iodine concentration [µg I (100 g FM)$^{-1}$]			
0		1.9 a		3.7 a	
0.1		39.9 b	75.6 b	51.9 b	55.6 b
0.2		167.2 cd	95.8 bc	107.1 c	104.6 c
0.5		328.9 e	296.0 de	186.6 cd	233.3 d
ONE WAY ANOVA		*** (0.000000)		*** (0.000000)	
GLM ANOVA two factorial	Block	NS (0.496869)		NS (0.554400)	
	Iodine form	NS (0.395612)		NS (0.383344)	
	Iodine dose	*** (0.000000)		*** (0.000000)	
	F x D	* (0.024357)		NS (0.593262)	

Figure 5.6 shows the postharvest treatment correlation (washed samples to unwashed samples) of different potted herbs. The strongly positive correlation (> 0.95) and the significant T-test (key to Figure 5.6) found in all cases suggests no influence due to the rinsing procedure and, consequently, indicating a complete uptake of the sprayed iodine solution through the wetted plant organs within a period of 5 - 14 days (cf. Figure 5.2).

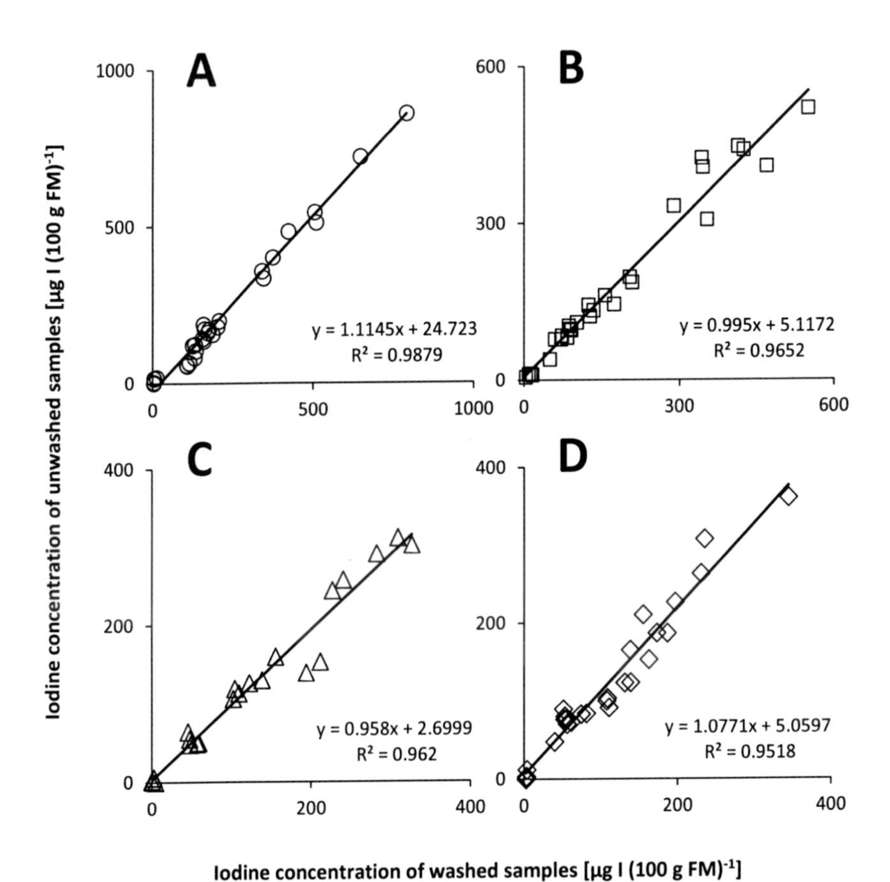

Figure 5.6 Correlation between washed and unwashed samples (postharvest treatments) of oregano [A, circles], basil [B, quadrates], parsley [C, triangles] and chives [D, diamonds] as affected by iodine foliar sprays at different concentrations (T-test [A]: probability level = 0.00, power = 1.00, T-value = 46.09, n = 28; T-test [B]: probability level = 0.00, power = 1.00, T-value = 26.86, n = 28; T-test [C]: probability level = 0.00, power = 1.00, T-value = 22.49, n = 22; T-test [D]: probability level = 0.00, power = 1.00, T-value = 22.21, n = 27). α = 0.001

5.4.3 Experiment no. 7: Miscibility of agrochemicals with iodine and influence of the application time and mode on the iodine accumulation behavior of butterhead lettuce

The miscibility of iodine with different agrochemicals and the influence of the application point in time on the iodine accumulation of vegetables were tested on butterhead lettuce sprayed at a dose of 0, (2x) 0.125 or (1x) 0.25 kg I ha^{-1}.

- **Tank mixing with calcium fertilizers**

The effects of iodate applied as a pure salt or as an emulsifiable concentrate (iodine fertilizer prototype) and combined with different commercial calcium fertilizers are shown in Figure 5.7. The marketable quality of butterhead lettuce was not affected and the biomass production showed no significant differences (Figure 5.7 A). Likewise, no significant differences regarding the calcium concentration in the dry matter fraction of butterhead lettuce could be observed after treatments with or without calcium containing fertilizers (Figure 5.7 B). The iodine concentration as affected by the same treatments is displayed in Figure 5.7 C. No iodine accumulation above the naturally occurring iodine content was found in the unfortified control and in the straight calcium fertilizer treatments. The straight iodate fertilizers sprayed as an EC or as a pure salt showed the same accumulation levels as when mixed with $Ca(NO_3)_2$. Remarkably, both iodate fertilizers showed an enhanced iodine accumulation when mixed with $CaCl_2$. Statistical significance was computed for the factors iodine and calcium fertilization and their respective interaction (key to Figure 5.7).

- **Tank mixing with pesticides**

The effects of KI and KIO_3 treatments combined with commercial pesticides are illustrated in Figure 5.8. Again, no visible or statistically significant differences were observed in the biomass production of butterhead lettuce (Figure 5.8 A). The iodine concentration in the edible plant parts, as affected by the same treatments, is displayed in Figure 5.8 B. Iodine containing treatments were significantly different throughout compared to the control and straight pesticide treatments. Although a slightly enhanced iodine concentration was observed in blended KI treatments, no significant differences between all the iodine containing treatments were found. Significance was computed for the factors iodine fertilization, pesticides and their respective interaction (key to Figure 5.8).

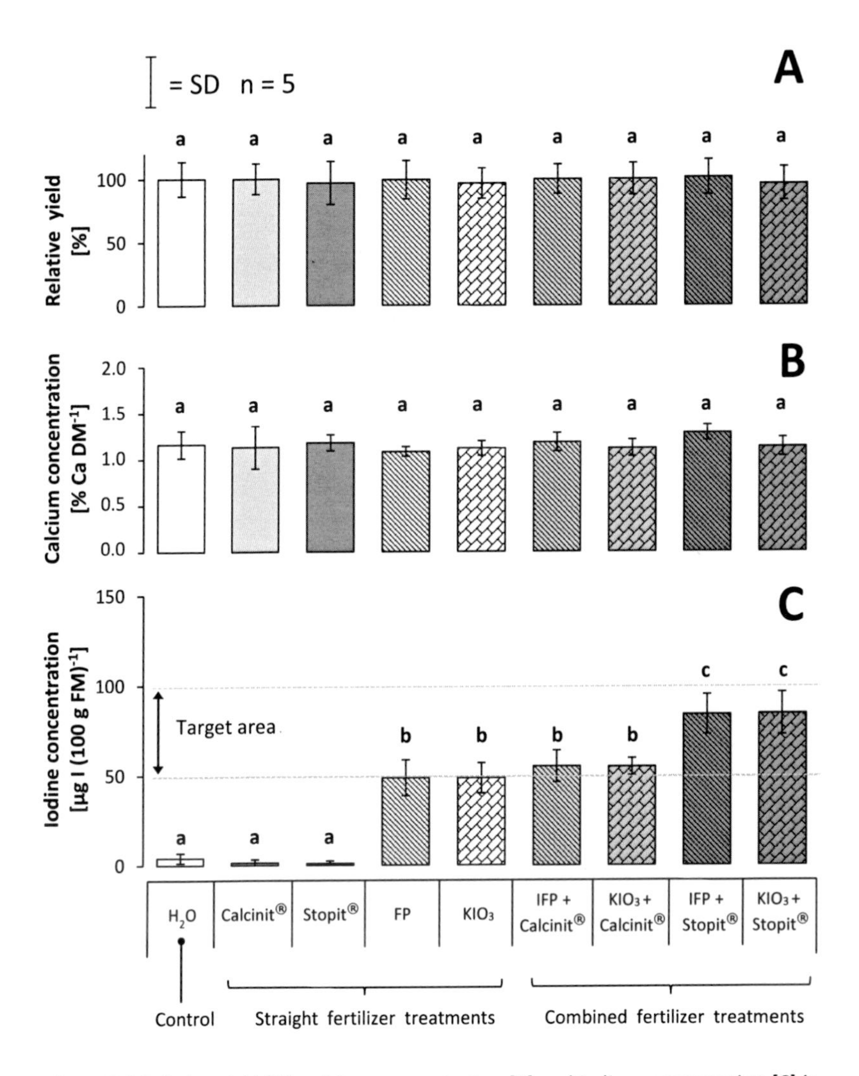

Figure 5.7 Relative yield [A], calcium concentration [B] and iodine concentration [C] in edible plant parts of butterhead lettuce as affected by iodine sprays (0 or 2x 0.125 kg IO_3^--I ha^{-1}) with or without commercial calcium fertilizers (0 or 2x 1 kg Ca ha^{-1}). All treatments sprayed at two different target dates with 0.02 % (v/v) Brake-Thru® surfactant. Treatments: H_2O = deionized water (0.05 µS cm^{-1}), Calcinit® = $Ca(NO_3)_2$, Stopit® = $CaCl_2$ (both YARA GmbH & Co. KG, Germany), IFP = fertilizer prototype (KIO_3 as EC), KIO_3 = iodate salt (pur. 99.8 %). Means with same letters do not differ according to Bonferroni MCP at α = 0.05 (One-way analysis of variance [C]: probability level = 0.00000, power = 1.00; GLM ANOVA [C]: iodine fertilization (I) p-level = 0.000001, calcium fertilization (CA) p-level = 0.000004, interaction I x CA p-level = 0.002712)

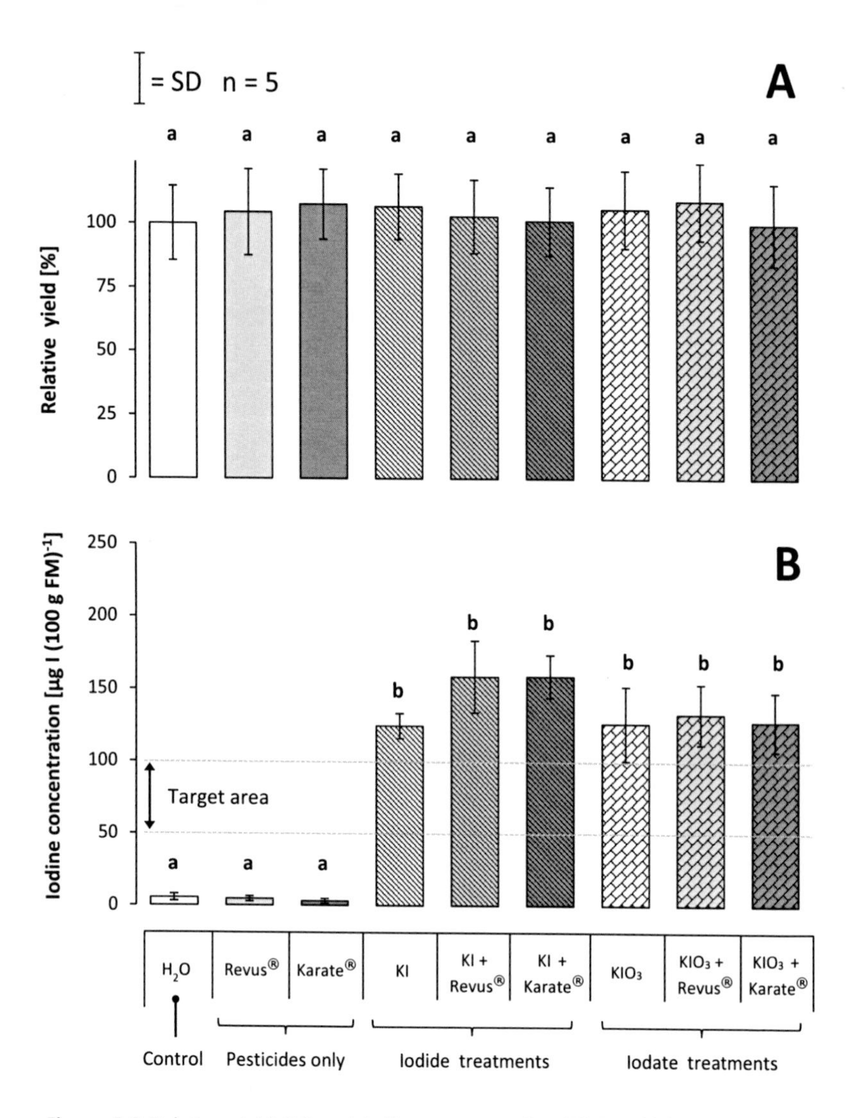

Figure 5.8 Relative yield [A] and iodine concentration [B] in edible plant parts of butterhead lettuce as affected by KI or KIO_3 sprays (0, 0.25 kg I ha^{-1}) with or without pesticides (Revus® at 600 ml ha^{-1} or Karate® at 75 ml ha^{-1}). All treatments sprayed with 0.02 % (v/v) Brake-Thru® surfactant at rate of 1000 L H_2O ha^{-1}. Means with same letters do not differ according to Bonferroni MCP at α = 0.05 (One-way analysis of variance [B]: probability level = 0.00000, power = 1.00; GLM ANOVA [B]: iodine fertilization (I) p-level = 0.049005, pesticides (P) p-level = 0.000000, interaction I x P p-level = 0.013638)

- **Diurnal application time**

The effects of iodine foliar applications (0 and 0.25 kg I ha^{-1} as KI or KIO_3) sprayed at different times of day are displayed in Figure 5.9. As in previous trials, no statistically significant differences were observed in the biomass production of butterhead lettuce (Figure 5.9 A). The unfortified control treatment indicates native iodine concentration levels (Figure 5.9 B). Remarkably, the KIO_3 treatments showed a lower but rather steady iodine accumulation during the course of the day, whereas the KI treatments revealed a significant peak at 11:00 and 15:00 hours. During this time span the insolation and the air temperature also peaked (bold values in the secondary X-axis of Figure 5.9 B). Comparing the treatments at 7:00 and 19:00 hours, a wider distinction between KI and KIO_3 was observed during the morning. At the same time the highest relative humidity (RH) value was recorded. Statistical significance was computed in this experiment for the factors iodine fertilization, application point in time and their interaction (key to Figure 5.9).

- **Pre-harvest intervals**

The influence of iodine sprays (0 and 0.25 kg I ha^{-1} as KI or KIO_3) applied at different growth stages of butterhead lettuce with a focus on the pre-harvest interval is shown in Figure 5.10. No visual or statistically significant differences were observed in the biomass production of butterhead lettuce (Figure 5.10 A). Higher iodine concentrations were found at decreasing pre-harvest intervals. Once more, the iodide treatments had a higher accumulation tendency than iodate treatments. The iodine concentration found in treatments sprayed 7 days before harvest was in a range close to or within the target area. The postharvest treatments (washed/unwashed) on butterhead lettuce samples sprayed one day before harvest showed a specific response depending on the iodine form: no differences were found in samples treated with KI, whereas KIO_3 treatments differed significantly.

The relation between the iodine content and the fresh mass as well as the leaf area of butterhead lettuce at different iodine application times before harvest are displayed in Figure 5.11 A and B. In the first case, the correlation was best described using a saturation curve whereas an exponential curve was most suitable for displaying the correlation to the leaf area. Figure 5.11 C shows data collected from different trials in experiment no. 7. Treatments were calculated as trial averages to evaluate the relation between iodine concentration and the relative growth of butterhead lettuce from the time of iodine application to harvest. Decreasing iodine concentrations were determined with increasing relative growth nearing harvest.

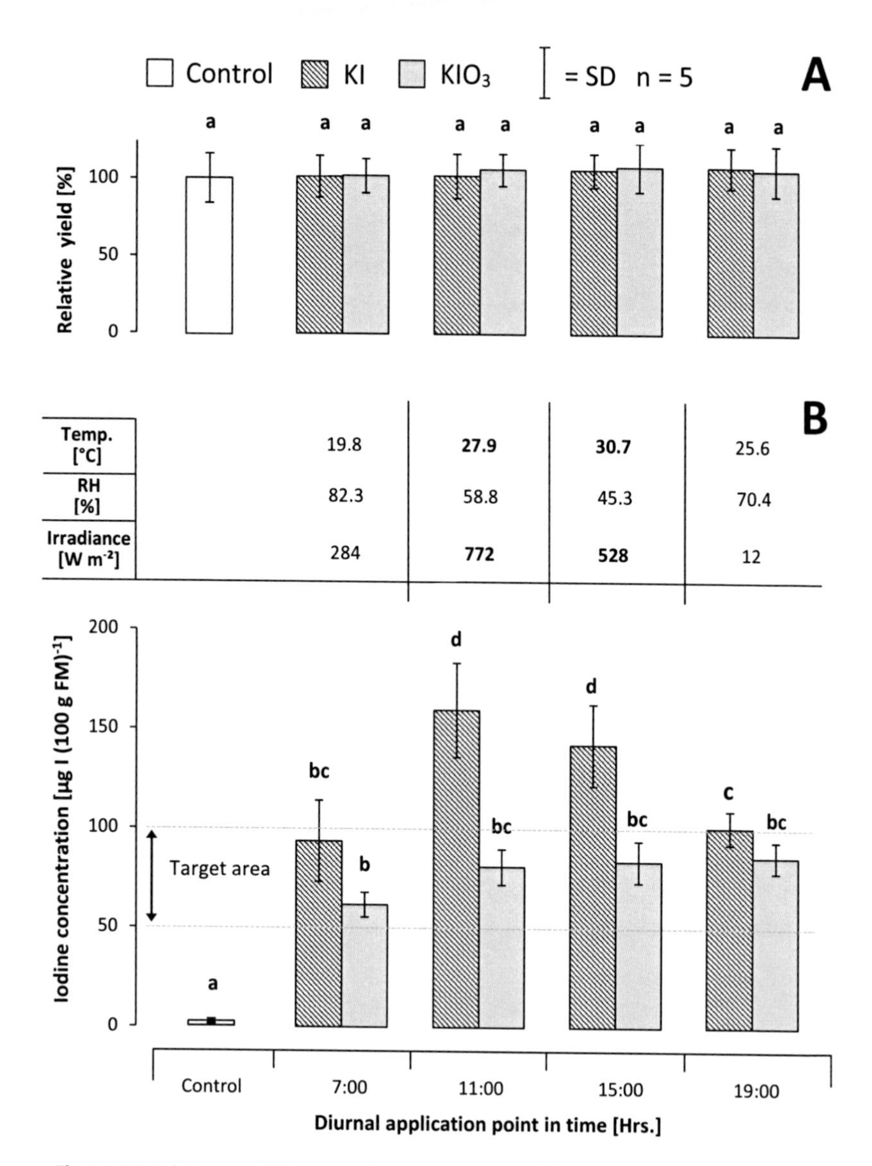

Figure 5.9 Relative yield [A] and iodine concentration [B] in edible plant parts of butter-head lettuce as affected by KI or KIO_3 sprays (0, 0.25 kg I ha^{-1}) at different application points in time during the day (control only at 11:00). Secondary X-axis shows temperature, RH and irradiance levels at the different application times. All treatments sprayed with 0.02 % (v/v) Brake-Thru® surfactant at rate of 1000 L H_2O ha^{-1}. Means with same letters do not differ according to Bonferroni MCP at $\alpha = 0.05$ (One-way analysis of variance [B]: probability level = 0.00000, power = 1.00; GLM ANOVA [B]: iodine fertilization (I) p-level = 0.000000, application time-point (T) p-level = 0.000014, interaction I x T p-level = 0.001041)

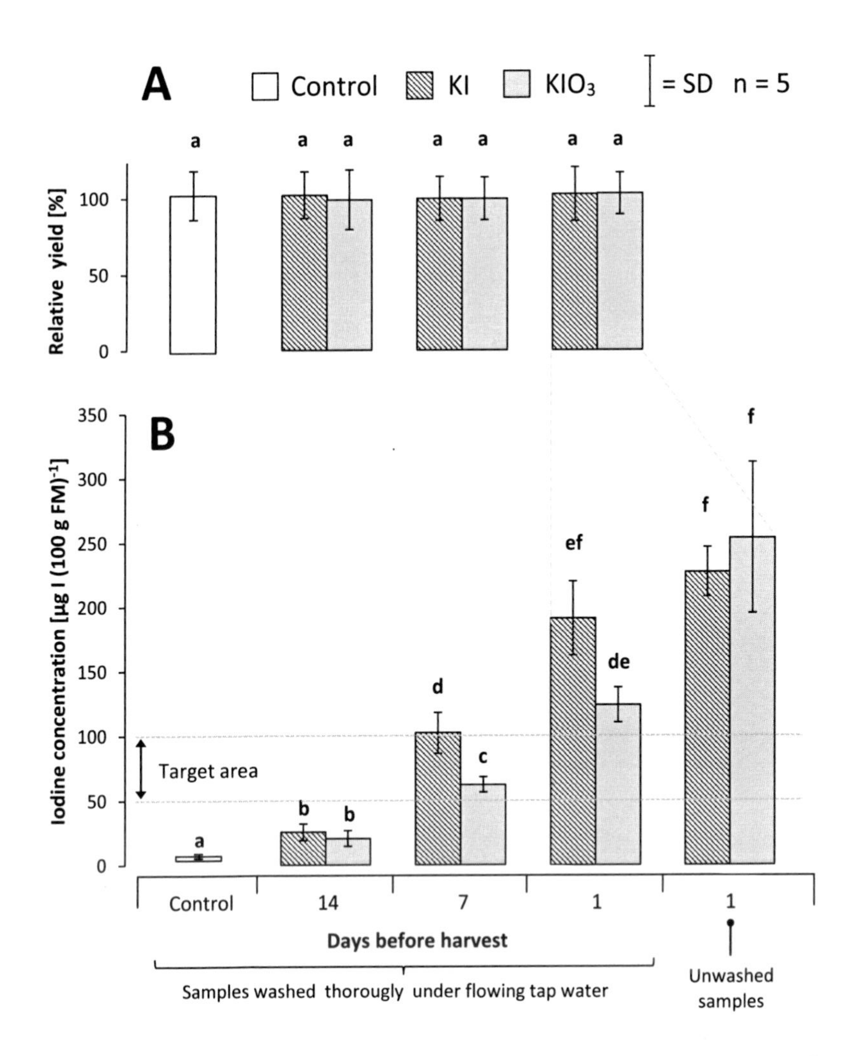

Figure 5.10 Relative yield [A] and iodine concentration [B] in edible plant parts of butterhead lettuce as affected by KI or KIO_3 sprays (0 or 0.25 kg I ha^{-1}) at different application points in time before harvest (control only 1 day before harvest). All treatments sprayed with 0.02 % (v/v) Brake-Thru® surfactant at rate of 1000 L H_2O ha^{-1}. Means with same letters do not differ according to Bonferroni MCP at $\alpha = 0.05$ (One-way analysis of variance [B]: probability level = 0.00000, power = 1.00; GLM ANOVA [B, two factorial]: iodine fertilization (I) p-level = 0.000076, application time-point (T) p-level = 0.000000, interaction I x T p-level = 0.351908 ; GLM ANOVA [B, two factorial]: iodine fertilization (I) p-level = 0.015212, post-harvest treatments (PT) p-level = 0.000009, interaction I x PT p-level = 0.001029)

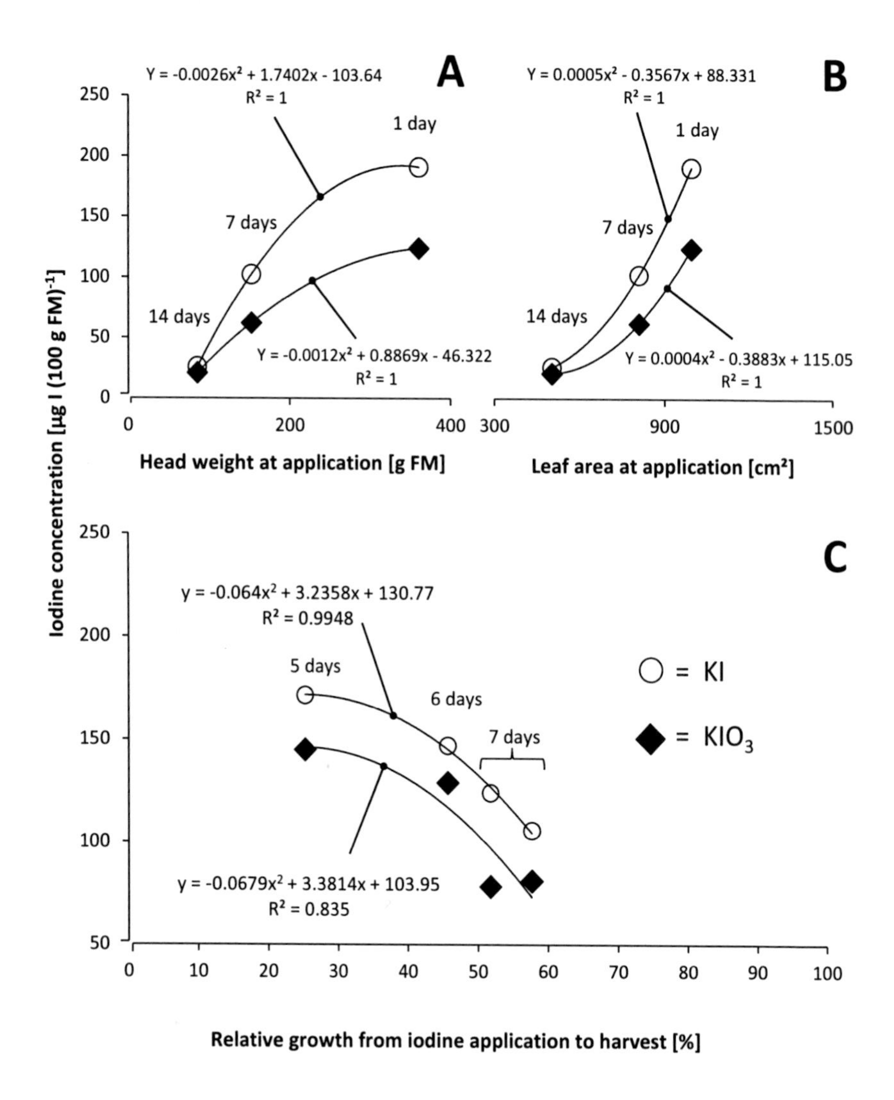

Figure 5.11 Relation between iodine concentration and fresh mass [A] or leaf area [B] of butterhead lettuce at different foliar spray application points in time; 14, 7 and 1 days before harvest, respectively. C = relation between iodine concentration and relative growth of butterhead lettuce from iodine application to harvest; 5, 6 and 7 days before harvest, respectively. Data collected from different trials in experiment no. 7 and calculated as trial averages. Empty circles = KI treatments; diamonds = KIO_3 treatments; all treatments at a dose of 0.25 kg I ha^{-1}

- **Water application rates**

The results of iodine foliar applications (0 and 0.25 kg I ha^{-1} as KI or KIO_3) sprayed at different water application rates (1000, 600 and 300 L H_2O ha^{-1}) are shown in Figure 5.12. No statistically significant differences were observed in the biomass production of butterhead lettuce as affected by different water application rates (Figure 5.12 A). Increasing iodine concentrations were found with decreasing water application rates and the accompanying increase of solute concentrations (Figure 5.12 B; secondary X-axis). The iodide treatments affected a higher accumulation tendency than iodate treatments. Statistical significance was computed for the factors iodine fertilization, water application rate and their respective interaction (key to Figure 5.12).

- **Comparison of application systems**

The effects of iodine foliar applications (0 and 0.25 kg I ha^{-1} as KI or KIO_3) sprayed by means of different application systems (common operational field sprayer vs. experimental hand held sprayer) are displayed in Figure 5.13. Although a slight decrease in biomass production of butterhead lettuce was noticed with the KIO_3 field-sprayer treatment, no significant differences were computed by the Bonferroni MCP. In contrast, the same dataset was found to be barely significant by the one-way ANOVA (key to of Figure 5.13 A). Iodine containing treatments were significantly different compared to the unfortified control (Figure 5.13 B). Treatments by means of mounted field-sprayer showed a more even iodine accumulation with no differences between the iodine form, whereas hand-held sprayer treatments created significant differences. However, statistical significance was computed for the factors iodine fertilization, application system and their respective interaction (key to Figure 5.13 B). It should be noted, that the relative application system efficiency (secondary Y-axis of Figure 5.13 B), calculated as average of the respective application system, is more than doubled in the experimental spraying system.

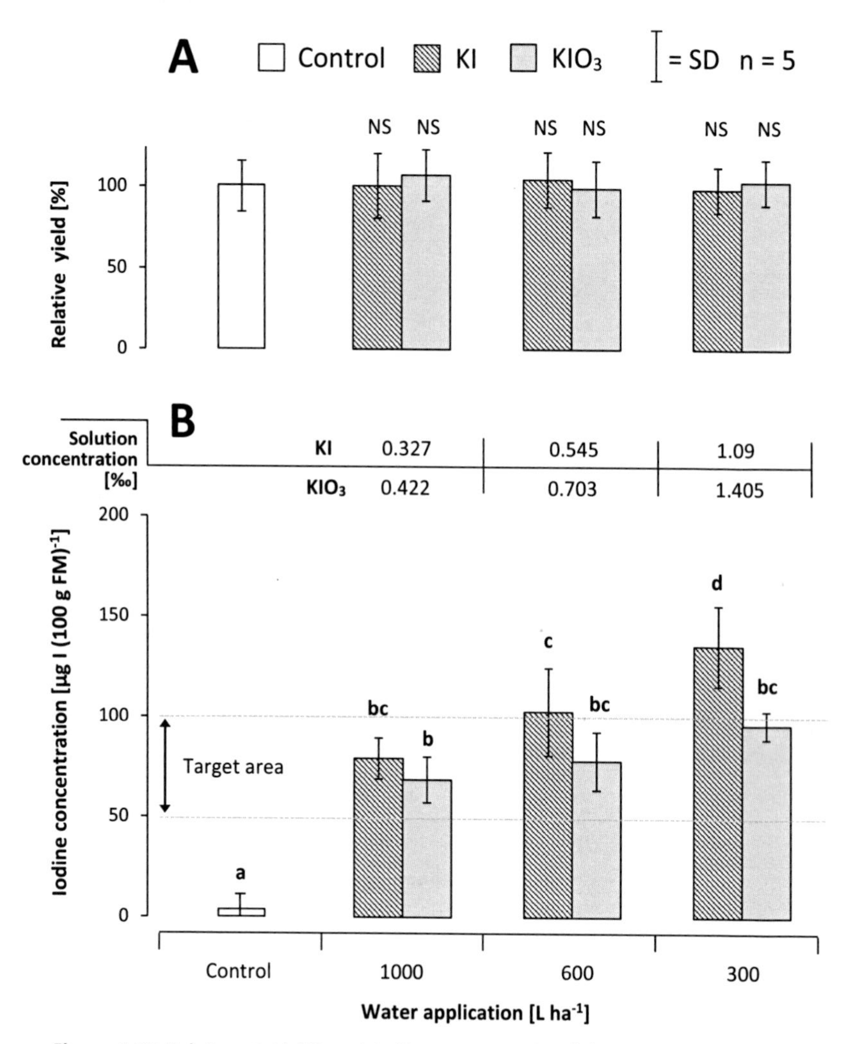

Figure 5.12 Relative yield [A] and iodine concentration [B] in edible plant parts of butterhead lettuce as affected by KI or KIO_3 sprays (0 and 0.25 kg I ha^{-1}) at different H_2O application rates (1000, 600, 300 L H_2O ha^{-1}; control only at 300 L H_2O ha^{-1}). Secondary X-axis shows KI and KIO_3 solution concentrations (‰). All treatments sprayed with 0.02 % (v/v) Brake-Thru® surfactant. Dataset A was compared to unfortified control by Wilcoxon rank-sum test. Levels of significance are represented by * = $p < 0.05$ and NS = not significant. Means of diagram B with same letters do not differ according to Bonferroni MCP at $\alpha = 0.05$ (One-way analysis of variance [B]: probability level = 0.00, power = 1.00; GLM ANOVA [B]: iodine fertilization (I) p-level = 0.001146, application rate (R) p-level = 0.000089, interaction I x R p-level = 0.167989)

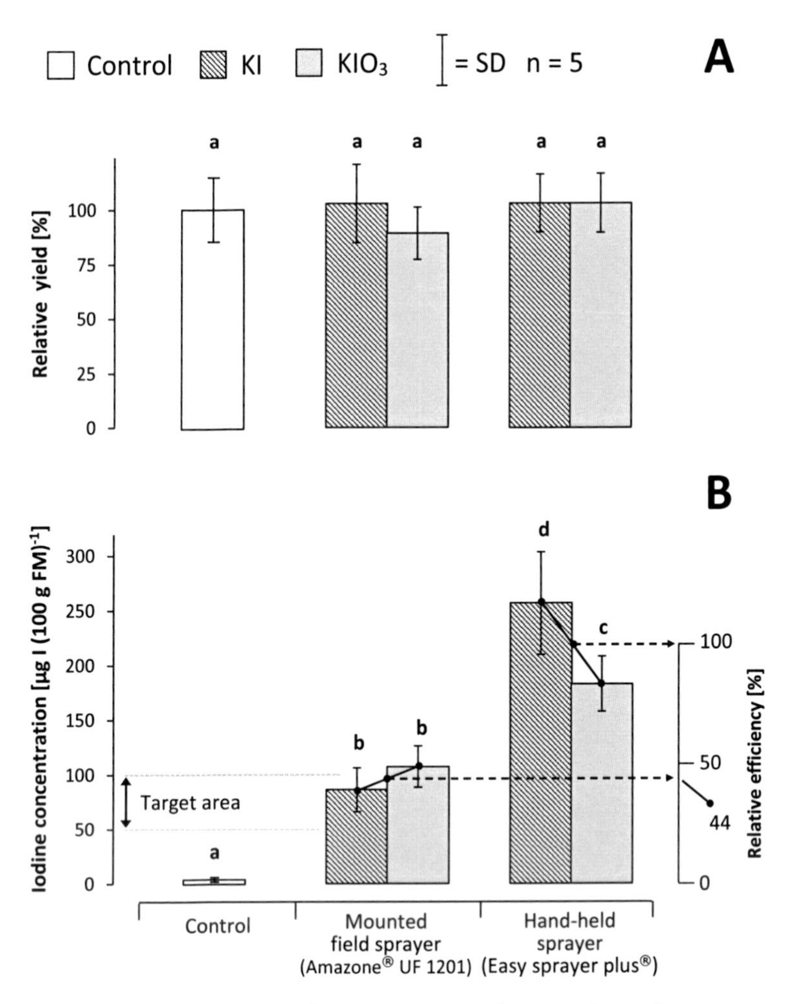

Figure 5.13 Relative yield [A] and iodine concentration [B] in edible plant parts of butterhead lettuce as affected by KI or KIO_3 sprays (0 and 0.25 kg I ha^{-1}) applied by common operational or experimental means (mounted field sprayer or hand-held sprayer). Secondary Y-axis shows efficiency averages of respective application system. All treatments sprayed with 0.02 % (v/v) Brake-Thru® surfactant at 1000 L H_2O ha^{-1}. Means with same letters do not differ according to Bonferroni MCP at α = 0.05 (One-way analysis of variance [A]: probability level = 0.041740, power = 0.696120; One-way analysis of variance [B]: probability level = 0.00000, power = 1.00; GLM ANOVA [B]: iodine fertilization (I) p-level = 0.000000, application system (AS) p-level = 0.000014, interaction I x AS p-level = 0.001041)

5.4.4 Experiment no. 8: Effects of iodine foliar sprays on the phenolic content of two multi-leaf lettuce varieties

The effects of different iodine foliar sprays (0 and 0.25 kg I ha^{-1} as KI or KIO_3) on a green and red Salanova® multi-leaf lettuce variety are displayed in Figure 5.14. Multi-leaf lettuce was found to tolerate the sprayed iodine dose of 0.25 kg I ha^{-1} well and no statistical differences in the biomass production were found by the parameter-free Wilcoxon rank-sum test compared to the unfortified control (Figure 5.14 A). Although the red variety had a lower fresh mass at the application point in time, differences in the relative growth, amounting to almost 20 %, were compensated for by the time of harvest. Significant differences between the two varieties were observed in the iodine concentration (Figure 5.14 B) and it is noteworthy that the red variety accumulated double the iodine quantity in the iodate treatment. Furthermore, the green variety showed significant differences between the iodine forms whereas the red variety displayed no distinction between them.

In Figure 5.14 C the concentration of total phenolic compounds as affected by iodine foliar sprays is shown. Almost double the concentrations were found throughout in the red variety. Expressed as percentages relative to the respective unfortified control, 6.2 % (I^- treatment) and 6.8 % (IO_3^- treatment) more total phenolic compounds were found in the green variety and 6.8 and 8.8 % more in the red variety respectively. Very narrow standard deviations were computed and significant differences to the control emerged. Compared to the iodine concentration, a remarkable inverse pattern between the iodine treatments was found: The response of the green variety made no distinction between the applied iodine forms, whereas significant differences were determined in the red variety. The analysis of variance of both iodine content and total phenolic concentration showed significant differences throughout for the factors variety, iodine form and their respective interactions (key to Figure 5.14).

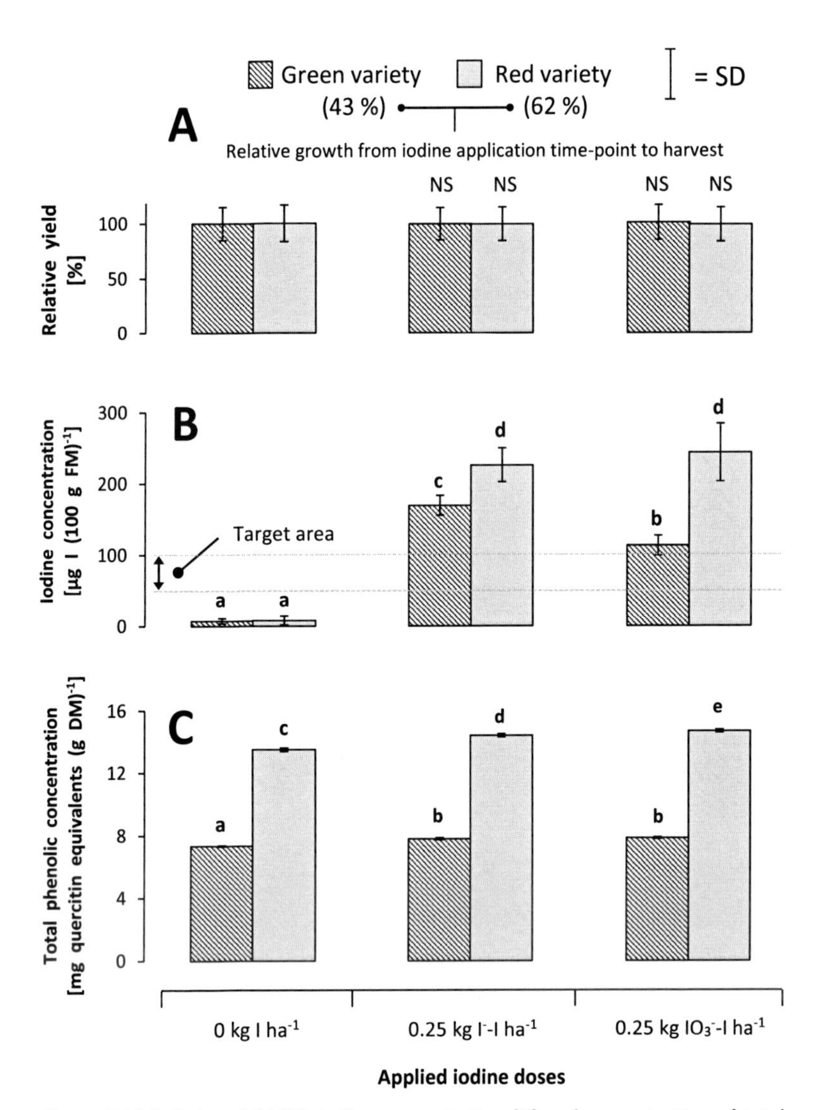

Figure 5.14 Relative yield [A], iodine concentration [B] and concentration of total phenolic compounds [C] in edible plant parts of Salanova® multi-leaf lettuce as affected by KI or KIO_3 sprays at 0 or 0.25 kg I ha^{-1}. All treatments sprayed with 0.02 % (v/v) Brake-Thru® surfactant at 1000 L H_2O ha^{-1}. Dataset A was compared to the respective control by Wilcoxon rank-sum test. Levels of significance are represented by * = $p < 0.05$ and NS = not significant. Means of B and C with same letters do not differ according to Bonferroni MCP at $\alpha = 0.05$ (One-way analysis of variance [B]: probability level = 0.00, power = 1.00; GLM ANOVA [B]: variety (V) p-level = 0.00, iodine form (I) p-level = 0.00, interaction V x I p-level = 0.00; One-way analysis of variance [C]: probability level = 0.00, power = 1.00; GLM ANOVA [C]: variety (V) p-level = 0.00, iodine form (I) p-level = 0.00, interaction V x I p-level = 0.000006). n = 5

5.4.5 Iodine distribution in different plant parts of butterhead lettuce and kohlrabi

The distribution of iodine in different plant parts of butterhead lettuce and kohlrabi as affected by single foliar sprays at 1 kg I ha^{-1} are displayed in Figure 5.15 (cf. experiment no. 4 in Table 4.2). In the case of butterhead lettuce the majority of iodine, up to six fold more was found in the inner part of the head than in the outer leaves (Figure 5.15 A). Remarkably high iodine concentrations were also found in the stalk of butterhead lettuce. In kohlrabi, a clear gradient was observed with the highest iodine concentrations in the leaf blades (Figure 5.15 B).

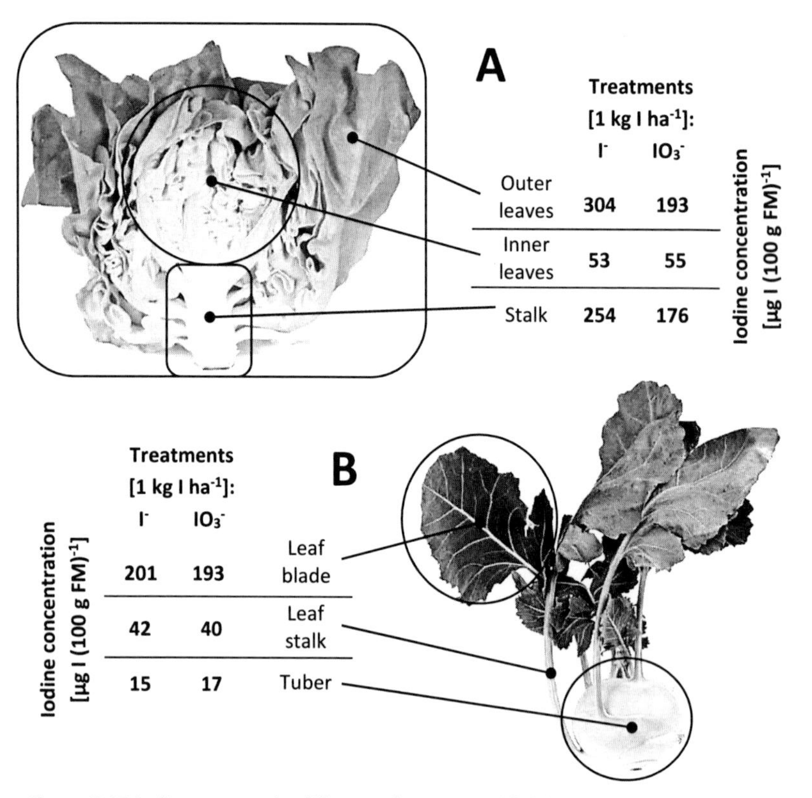

Figure 5.15 Iodine content in different plant parts of [A] butterhead lettuce and [B] kohlrabi as affected by KI or KIO_3 sprays at 1 kg I ha^{-1} sprayed twice 7 and 3 days before harvest, respectively

5.4.6 Leaf surface topography of some vegetable species

In the following, some examples are given for the observations made in preliminary investigations on the leaf surface topography of different vegetable species by means of light microscopy (LM) and scanning electron microscopy (SEM). Although semi-quantified in some cases (cf. 5.3.3 and key to Figure 5.17), the results in this section will only be described as tendencies or trends; this is mainly due to the lack of micrographs in sufficient quality and number to allow statistical analysis.

The stomata sizes of winter spinach, butterhead lettuce and white cabbage are shown in Figure 5.16. Large to very large stomata with a pronounced elliptical shape were detected in butterhead lettuce and winter spinach. In contrast, white cabbage had markedly smaller, more round to oval stomata. The stomata density of these species is shown in Figure 5.17. Winter spinach and white cabbage had a comparable density (> 100 Stomata mm^{-2}) whereas butterhead lettuce showed the lesser number of stomata per units of area (< 50 Stomata mm^{-2}).

The stomata comparison between red and green multi-leaf lettuce is depicted in Figure 5.18. Surprisingly, the green variety showed stomata of a smaller size (Figure 5.18 A compared to B) but generally a larger number of stomata per unit of area compared to the red variety (Figure 5.18 C compared to D).

In Figure 5.19 and 5.20 characteristical trichomes, i.e. fine epidermal outgrowths of various kinds, are depicted using the example of apparently glabrous (butterhead lettuce) and evidently pubescent leaves (oregano). In butterhead lettuce, the trichomes were oblong, roundly shaped, ending with a roundly-headed tip. The trichomes encountered on the leaf surface of oregano were remarkably different in morphology (glandular and hair-like trichomes; Figure 5.20 A), density and size. Glandular trichomes were round, shallow and separated from the adjacent cells by a fissure. In contrast, hair-like trichomes were long and acuminate with adjacent epidermal cells.

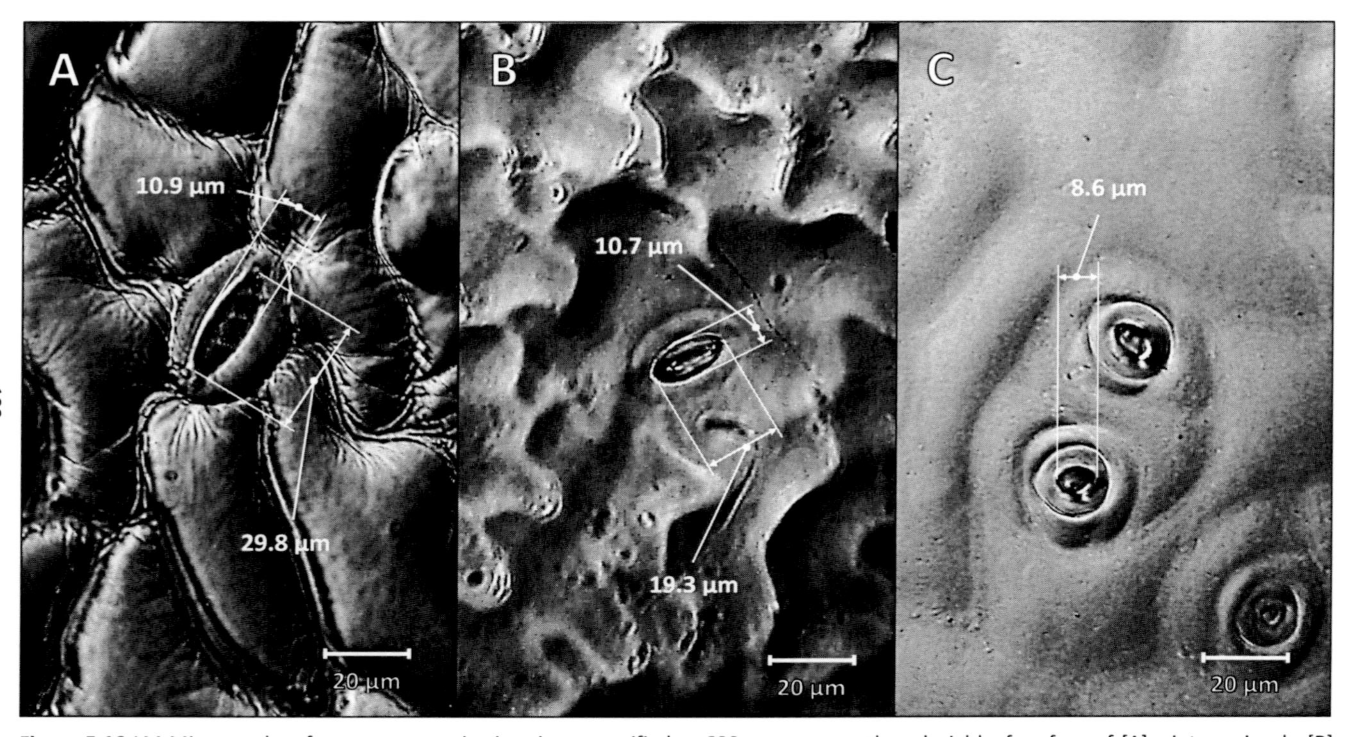

Figure 5.16 LM-Micrographs of stomata pore-size imprints magnified at 600x present on the adaxial leaf surface of [A] winter spinach, [B] butterhead lettuce and [C] white cabbage

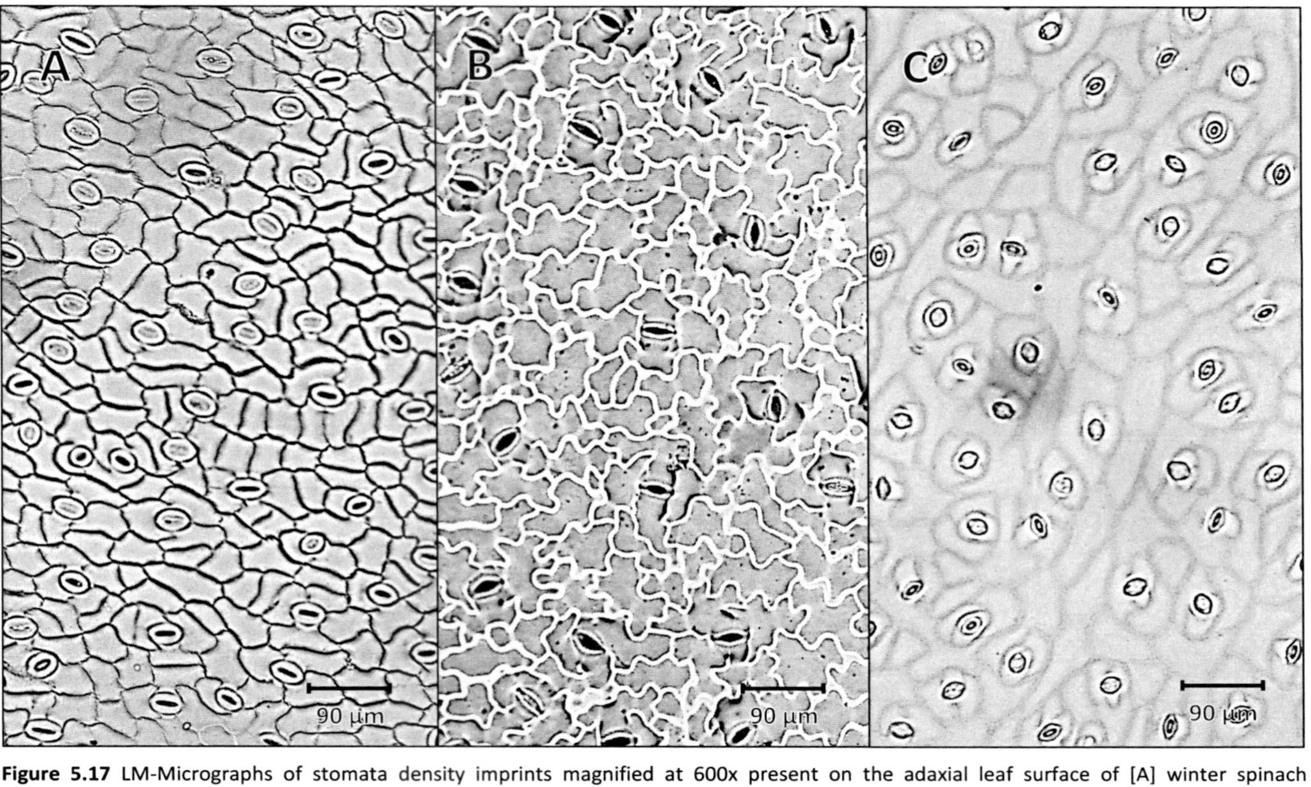

Figure 5.17 LM-Micrographs of stomata density imprints magnified at 600x present on the adaxial leaf surface of [A] winter spinach (> 100 stomata mm^{-2}), [B] butterhead lettuce (< 50 stomata mm^{-2}) and [C] white cabbage (> 100 stomata mm^{-2})

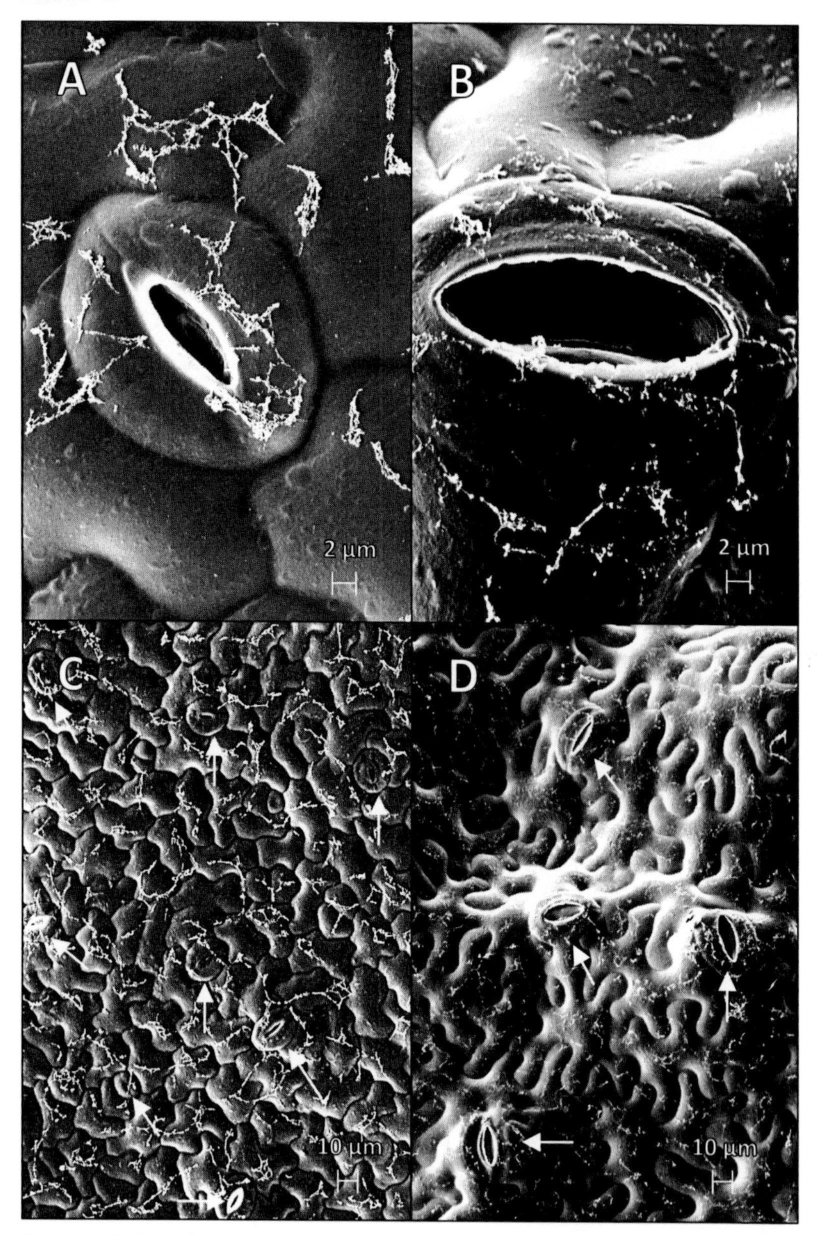

Figure 5.18 Scanning electron micrographs of stomata present on the adaxial leaf surface of Salanova® multi-leaf lettuce cv. `Archimedes RZ´ (green variety) magnified at 400x [A] and 2350x [C] and Salanova® multi-leaf lettuce cv. `Gaugin RZ´ (red variety) magnified at 400x [B] and 2350x [D]. White arrows indicate stomata

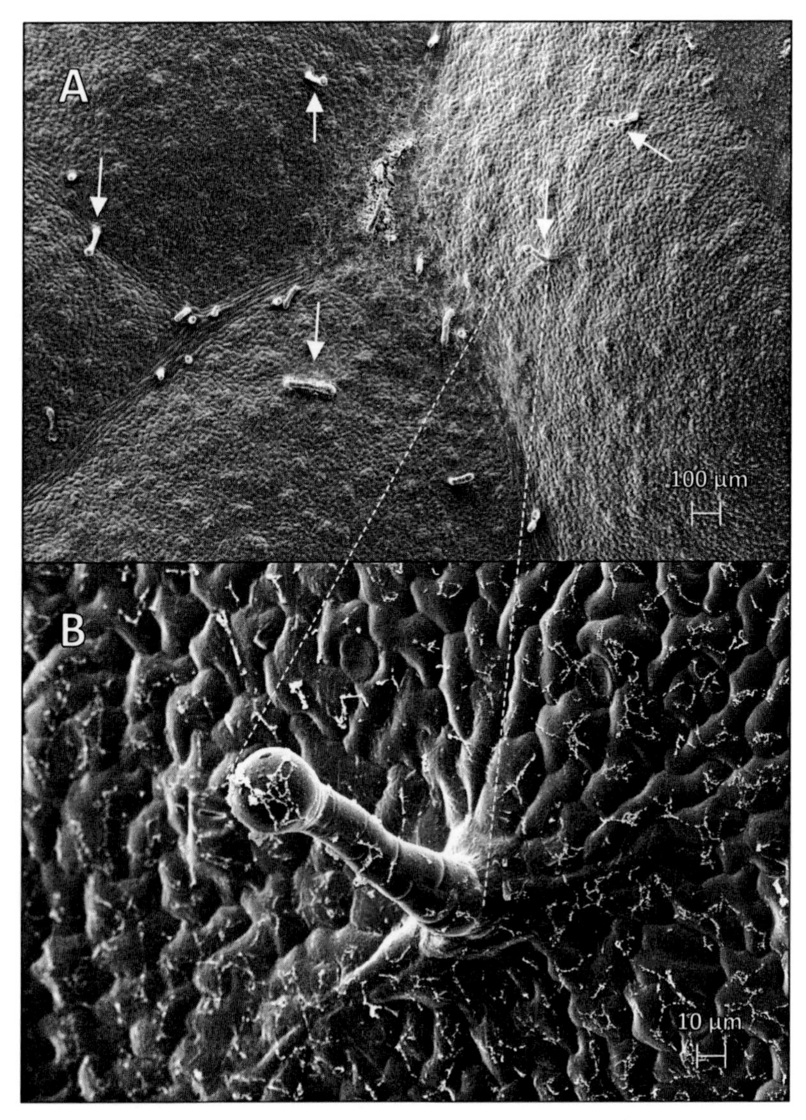

Figure 5.19 Scanning electron micrographs of trichomes magnified at 40x [A] and 400x [B] present on the adaxial leaf surface of butterhead lettuce. White arrows indicate dispersed trichomes

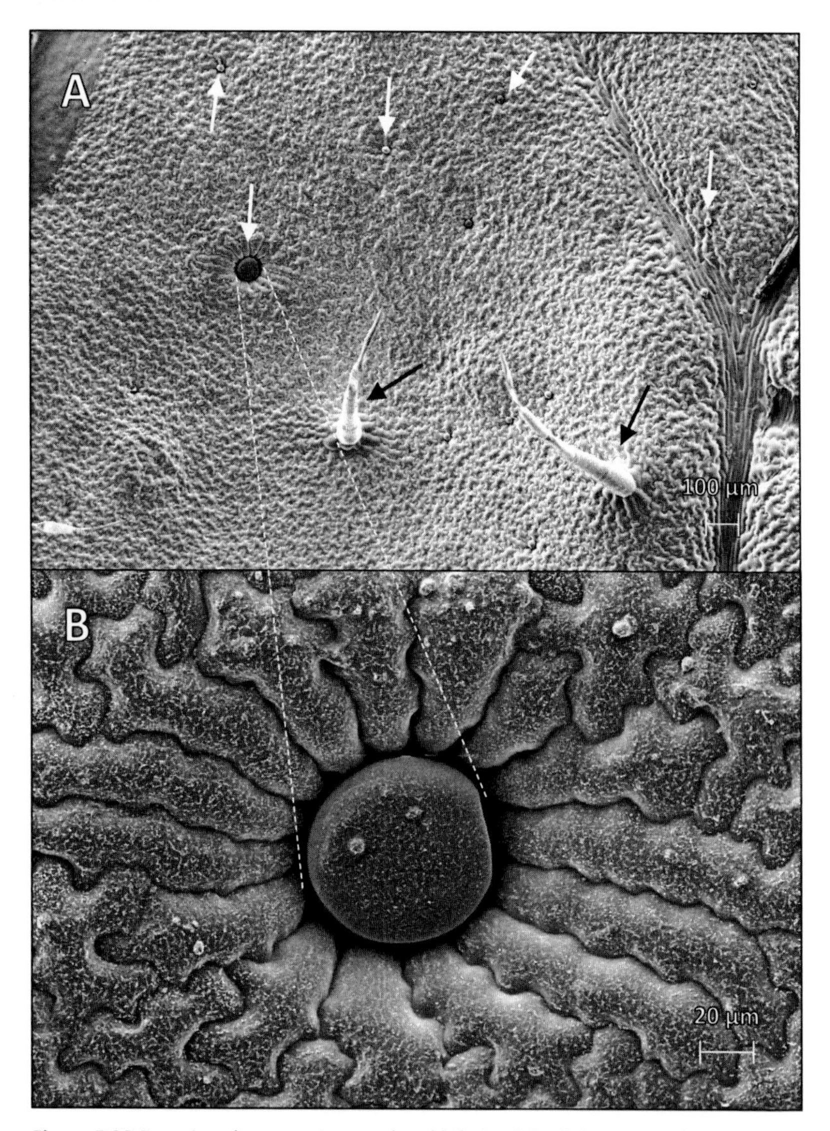

Figure 5.20 Scanning electron micrographs of [A] glandular (white arrows) and hair-like trichomes (black arrows) magnified at 50x and [B] a single glandular trichome magnified at 400x present on the adaxial leaf surface of oregano

5.5 Discussion

5.5.1 Crop yield and marketable quality as affected by iodine foliar fertilization

The use of single or twofold foliar sprays, in the range of 0.1 - 1.0 kg I ha^{-1}, was found to be well tolerated by the most vegetable species investigated in this thesis. Nevertheless, a biomass increase as reported by ALTINOK ET AL. (2003) could not be observed in any of the cases and thus our results are consistent with observations made by STRZETELSKY ET AL. (2010) and SMOLEŃ ET AL. (2011a, 2011b). On winter spinach and crisp lettuce severe leaf burn was ascertained at doses in the range of 0.5 to 1 kg I ha^{-1} applied as KI or KIO_3 (cf. Figure 5.4 and 5.5) and, in crisp lettuce, a significant yield depression was computed (cf. Table 5.4). Diverse significances in statistical computation of crop yield between winter spinach and crisp lettuce may have resulted as a consequence of the different crop developmental stages at the iodine application points in time: Winter spinach was sprayed at a growth stage close to maturity whereas the crisp lettuce development was impaired at a middle growing stage (cf. Figure 5.1 E). The recovery time and thus the lost growth advantage, in comparison to the unfortified control, probably led to the significant yield reduction only in the crisp lettuce.

The observations on detrimental effects of iodine foliar sprays are partially in concordance with findings of STRZETELSKY ET AL. (2010) who reported yield depression by spraying KI on radish at a dose of 2x 0.8 kg I ha^{-1}. KIO_3 treatments at the same dose did not cause any impairment. In contrast, KIO_3 treatments on crisp lettuce were found to be more detrimental than KI treatments whereas winter spinach seemed to react in a similarly susceptible way to both KI and KIO_3 treatments. Genotypic differences probably led to a specific response pattern and some vegetable crops thus seem to react more sensitively to iodine foliar sprays than others. Specific differences in the leaf morphology and structure may further explain these observations. For susceptible species, a modified fertilization strategy with repeated treatments at low iodine doses may lead to the desired iodine accumulation without leaf burn.

5.5.2 Iodine accumulation in different vegetable crops as affected by foliar sprays at varied iodine doses

The investigated vegetable species could be separated into good (butterhead lettuce and rocket), poor (white cabbage, broccoli and inner leaves of crisp lettuce) and excessive (winter spinach and outer leaves of crisp lettuce) iodine accumulators (cf. Table 5.5). Hence, biofortification strategies must be adapted specifically with the focus on the morphology and the general accumulation behavior of the examined species.

In butterhead lettuce and rocket, the iodine content could be enhanced to a satisfactory amount; iodine doses of 0.5 kg I ha^{-1} were sufficient or already too high to achieve the desired iodine concentration in edible plant parts. Nevertheless, a large variability in the iodine accumulation behavior of three successive butterhead lettuce crops cultivated and sprayed at different dates during the year (cf. Figure 5.1) suggest a high influence of exogenous factors like insolation, temperature, relative humidity or irrigation. Moreover, the continual comparison between the two iodine forms for the different vegetable species offered no valuable clue about the more appropriate form to spray. Trials focusing on the effects of different application factors are discussed in detail below (cf. 5.5.4).

White cabbage and broccoli showed the least accentuated iodine absorbability with levels far below the target area. Possible explanations for this observation are an unfavorable leaf area to fresh mass ratio, dilutory effects due to the large fresh mass of white cabbage heads (> 2.5 kg) and general characteristics of the cabbage leaf structure. The waxy cuticular layer of plants of the cabbage family (*Brassicacea*) may be one of the key factors suppressing the iodine uptake: ENNIS ET AL. (1952) reported a high repellence of water droplets on cabbage and an increased retention of the spray solution by addition of varied surfactants. More recently, RUEEGG ET AL. (2006) described an increased efficacy of 20 %, while spraying pesticides on broccoli, by adding 0.05 % (v/v) of the same adjuvant used in our trials (Brake-Thru® S 240). In contrast, KOCH (2007) reported an improved, neutral or even reduced pesticide efficacy, depending on the commercial pesticide formulation, by adding Brake-Thru® to the foliar spray solutions applied on rapeseed (*Brassica napus* L.). The epicuticular waxes of hydrophobic leaves, like those of cabbage, were indeed found to greatly affect the wetting behavior of different surfactants (ZHANG ET AL. 2006). Moreover, the effectiveness of aerially applied mineral nutrients (N, P and K applied on leafy cabbage, collard greens and kale) has been regarded as questionable since poor accumulation rates have been observed: the period of time for the uptake extended up to 24 hours for nitrogen and up to 4 and 15 days for potassium and phosphorous, respectively (STRANG ET AL. 1997).

A further explanation for the minimal iodine content in the edible plant parts of white cabbage and broccoli may be found by considering the well documented inhibitory effect of the thiocyanate anion (SCN^-) on the absorption and transport of iodide in humans: In the thyroid gland, SCN^- acts as an antagonist since it is a monovalent anion similar in molecular size to iodide (ERDOGAN 2003). The genus *Brassica* is, in turn, known to produce large quantities of thiocyanates (JENSEN ET AL. 1953; GREER 1957; STOEWSAND 1995). It is thus conceivable that once absorbed by leaves, the translocation of I^- in plants may be interfered with by the concurrent effect of SCN^-. However, this is mere speculation since no further supporting references could be found.

The phytotoxic effects observed in winter spinach and outer crisp lettuce leaves suggest high iodine accumulation rates for both vegetable species. The morphology of crisp lettuce may in turn explain the low iodine accumulation levels found in the edible plant parts: The formation of a close head with leaf layers wrapped around a common core, which are separated by air cushion layers, may have impeded the penetration and diffusion of iodine into the interior of the head. Furthermore, these observations suggest a rather poor phloem mobility of aerially applied iodine (details on the phloem mobility of iodine will be discussed in section 5.5.6). However, a biofortification strategy at a reduced dose up to a total of 0.05 - 0.1 kg I ha^{-1} for winter spinach and multiple sprays during the cultivation period for crisp lettuce may lead to satisfactory results without leaf impairment and should be tested in future trials.

While iodine hyper-accumulators can probably be biofortified at a reduced iodine dose, it should be stated that some crops, e.g. the genus *Brassica*, may be inadequate for iodine biofortification since low operating costs are decisive for a practicable fertilization procedure. Nevertheless, the addition of different spray adjuvants such as penetrators (oils), humectants or synergists to the solution and multiple iodine applications at doses of $\geq$ 1 kg I ha^{-1} from the very beginning of the cultivation, may lead to a satisfactory iodine accumulation in some cabbage varieties.

Generally, the comparison between pure grade KIO_3 and KIO_3 as an emulsifiable concentrate [see late planting (III) in Table 5.5] showed slightly higher iodine concentrations using the iodine fertilizer prototype and a statistically significant difference at the 0.75 kg IO_3^--I ha^{-1} treatment. These observations may be attributed to an improved formulation of the EC, but further trials are needed to validate the advantage of the tested iodine fertilizer prototype.

5.5.3 Iodine accumulation behavior of some culinary herbs cultivated in greenhouses

The iodine content of the investigated potted herbs species could be enhanced to a satisfactory amount at the lowest application dose of 0.1 kg I ha^{-1}. Since the fresh mass amounts of herbs used as food ingredient are usually far below the reference value of 100 g FM, higher iodine application doses up to 0.5 kg I ha^{-1} are supposably uncritical, assuming a portion of a few grams fresh herb weight per meal.

Moderate to broad differences in the iodine accumulation, again with no clearly dominant iodine form, could be observed in the investigated plant species. The descending order of iodine accumulation was: oregano > basil ≥ parsley > chives. The different morphology, for instance the plant canopy, may partially explain these results: Although having only tiny leaves, oregano has very dense foliage that leads to a specific leaf area of 40.6 m^2 (kg DM)$^{-1}$ (Poorter and Remkes 1990). Chang et al. (2008) reported an approximately 25 % smaller specific leaf area in basil [control treatment = 35.2 m^2 (kg DM)$^{-1}$]. Although without quantification values for the leaf area, chives was observed to have only very short scapes (4 - 8 cm; 2 days after propagation) at the application time and, consequently, probably a smaller surface area compared to the other species. Additionally, the relative growth of the different herb species from application time to harvest may have influenced the iodine concentration to a greater or lesser extent with dilutory effects. Additionally, other specific morphological and structural components, e.g. trichomes, stomata, cuticular cracks or imperfections, which are important uptake pathways for solutions on the leaf surface (Fernández et al. 2013), may have played a decisive role in the iodine uptake of the investigated herbs (leaf topography is discussed in detail in section 5.5.7).

The successful uptake at a high water application rate of 4,000 L H_2O ha^{-1} makes overhead fertigation with irrigation booms or gantries a cost effective and labor saving proposition. However, since water application rates are regulated specifically depending on the needs of the grower, the cultivated species and environmental factors, the fertigation under varying practice conditions should be tested and evaluated considering the above-mentioned factors.

5.5.4 Miscibility of agrochemicals with iodine and influence of the application time and mode on the iodine accumulation behavior of butterhead lettuce

The combination of more than one agrochemical compound in so-called tank mixes is a cost-effective and common practice among growers (GRIFFITH 2010). A limited range of compounds was tested (cf. 5.4.3) using butterhead lettuce as a model system. The miscibility of iodine is regarded as a crucial factor for iodine biofortification since operational costs are a limiting factor in the application. Other decisive aspects for practical fertilization are the influence of fluctuating environmental conditions on iodine foliar sprays, the relative growth stage at the application time and varied water application rates. Finally, the comparability of the experimental application to the common operational system (small scale plot trials vs. large scale application) will be decisive for the transfer of trial results to practical use.

- **Tank mixing with calcium fertilizers**

Pure grade KIO_3 and the KIO_3 fertilizer prototype were mixed with the commercial calcium fertilizers Calcinit® [$Ca(NO_3)_2$] and Stopit® ($CaCl_2$; cf. Figure 5.7). Calcium levels in the dry matter fraction remained unaffected by calcium sprays. This is not unusual considering that the maximum uptake of calcium was limited to 28.57 mg Ca plant^{-1}, which corresponds to approx. 0.2 % Ca DM^{-1} (calculation data: total calcium application = 2 kg Ca ha^{-1}; crop density = 70,000 plants ha^{-1}; mean head weight = 350 g; dry matter = 4 %). Furthermore, poor uptake rates for aerially applied calcium have been reported in literature and explained as being due to low calcium concentrations in the application solution (0.1 % Ca) and a low relative humidity (approx. 60 %) at the time of application (SCHÖNHERR 2000). The calcium accumulation levels in our trials were consistent with results reported by others who tried to prevent tip burn with calcium sprays on different lettuce varieties (ABD EL-FATTAH AND AGWAH 1987; HOLTSCHULZE 2005).

The satisfactory iodine accumulation achieved by mixing iodine with commercial calcium fertilizers proved no negative influence of Calcinit® and a positive influence of Stopit® on the iodine foliar sprays. An increased iodine accumulation in the latter case can be explained by the influence of several adjuvants in a fully formulated commercial product or by taking the humectant properties of $CaCl_2$ itself in account. Humectants in solutions have positive properties on foliar applications, for example by lowering the point of deliquescence (POD) of the solution and the consequential prolongation of the drying process on the leaf surface as well as by increasing the retention rate and the rain fastness (SCHMITZ-EIBERGER ET AL. 2002; KRAEMER ET AL. 2009; BLANCO ET AL. 2010; FERNÁNDEZ ET AL. 2013). The most plausible explanation for the enhanced iodine accumulation is the sum of all the positive effects of the adjuvants and the main components and probably their mutual reaction. Since no information is available on the ingredients of Stopit®, further miscibility trials must be conducted to elucidate the effects of pure grade $CaCl_2$ as the sole adjuvant in an iodine solution.

The comparison between pure grade KIO_3 and KIO_3 as an emulsifiable concentrate showed no statistically significant differences. Differences observed in a previous trial [see late planting (III) in Table 5.5] might have been evened out by further reducing the iodine application dose to 0.25 kg IO_3^--I ha^{-1}: Adjuvants present in the EC have been diluted to the same extent as the iodine doses have been diminished, whereas the added Brake-Thru® nonionic wetter in the pure grade KIO_3 solution maintained the constant concentration of 0.02 % (v/v). However, since no information on the ingredients and their respective quantities is available for the fertilizer prototype, the only conclusion that can be made is that the effectiveness of the tested formulation is heavily dependent on the composition and concentration of the surfactants present in the fertilizer solution (STOCK AND HALLOWAY 1993). The same iodine doses and similar adjuvants and adjuvant doses in the spray solution will probably lead to comparable results.

- **Tank mixing with pesticides**

No detrimental effects were observable when mixing the insecticide Karate® or the fungicide Revus® with pure grade KI or KIO_3 salts. Even though no statistically significant differences were computed, slightly higher concentrations were observed mixing KI with the two pesticides (cf. Figure 5.8). Multiplicative effects between the adjuvants in the commercial formulation and potassium iodide could be the reason for a slightly increased iodine uptake.

These first miscibility trials proved that iodine is likely to tank mix well with commercial formulations. Although iodine is unlikely to negatively influence the effectiveness of the pesticides, in part because of the small application quantities, interferences cannot be completely excluded, especially when tank mixing is performed with a wide variety of different agrochemicals. For instance, the addition of the micronutrients boron and manganese to the pyrethroid insecticides fenpropathrin and lambda-cyhalothrin was found to dramatically change the solution pH (CHAHAL ET AL. 2012). Hence, multiple interactions of agrochemicals with iodine should be further examined in miscibility trials with a focus on the solution pH and the redox potential.

- **Diurnal application times**

The application of iodine salts on butterhead lettuce was tested under varying environmental conditions such as fluctuating insolation, relative humidity and temperature over the course of the day (cf. Figure 5.9). The iodine uptake with KIO_3 treatments remained largely unaffected by fluctuating conditions whereas KI treatments showed significant peaks at the highest light insolation and air temperature. Environmental factors such as light radiation and temperature or water status of the leaf are in turn known to directly influence the guard cells and to eventually trigger the opening of the stomatal pores (TAIZ AND ZEIGER 2006). Although all the mechanisms of solute movement into the foliar surface are not yet completely understood, stomata may play a dominant role in the uptake of aerially applied nutrients (FERNÁNDEZ ET AL. 2013). The process of stomatal uptake was demonstrated to occur by diffusion along the guard cell walls and to be slow and size selective (EICHERT AND GOLDBACH 2008). The diffusion coefficient in Fick's first law indicates how easily a substance moves through a medium. Higher temperatures and smaller molecules lead to a higher diffusion coefficient and hence to faster diffusion down a concentration gradient (TAIZ AND ZEIGER 2006). The intrinsic difference between KI and KIO_3 treatments can thus be interpreted by considering the ionic size and the specific hygroscopicity: The smaller iodide anion has a higher diffusion coefficient and the higher hygroscopicity of KI (PAHUJA ET AL. 1993; DIOSADY ET AL. 2002) leads to a longer drying time on the leaf surface. This might provide a significant advantage in the uptake process. Hygroscopic substances have a specific point of deliquescence (POD) that has been defined as the relative humidity value, at a given temperature, were the liquefaction of a salt into a solute takes place (GREENSPAN 1977; SCHÖNHERR 2001; FERNÁNDEZ AND EICHERT 2009). The points of deliquescence for KI and KIO_3 are 68.9 and 93.8 % RH at 20 °C, respectively (GREENSPAN 1977; APELBLAT AND KORIN 1998). After a short drying period, generally only a few minutes depending on the RH at the time of application, non-absorbed solutes crystallize on the leaf surface. It is thus conceivable that a complementary uptake of iodine salts may have occurred at a later point in time, for instance in the course of the following night(s) after application during a higher RH phase. Hence, the lower POD of KI may have presented a further advantage for the foliar uptake at a later date. On the other hand, KIO_3 may have superior practical properties, being less sensitive to environmental fluctuations. This is especially the case when considering the feasible iodine accumulation predictability and a more stable product with a prolonged shelf life (PAHUJA ET AL. 1993). However, humidity, temperature and light conditions seems to have multiple effects on the iodine uptake of leaves and are decisive for the inherent efficiency differences between the two iodine forms. Therefore, further trials under regulable ambient conditions in climate chambers should be conducted to better understand the complex influence and interaction of these factors.

- **Pre-harvest intervals**

The iodine accumulation behavior of butterhead lettuce was tested with regard to the pre-harvest interval and the consequential different growth stages of the plant at the time of iodine application (cf. Figure 5.10). An increased iodine accumulation was observed with increasing fresh mass and leaf area, particularly using KI (cf. Figure 5.11 A and B). However, the relationship of the iodine concentration to the leaf area was calculated on basis of two-dimensional image data which tends to overestimate the surface area influence by offering smaller increments than actually present. The real increases in the total surface area of a growing lettuce head would be significantly larger and the observed exponential relationship would probably turn into a saturation curve similar to the one of the fresh mass relationship. Hence, the relationship of the iodine concentration to the different growth stages can be described more accurately by using the fresh weight that reaches its full saturation close to the harvest. In addition, the evaluation of leaf area data would not be feasible during the field inspection of crops whereas a spring scale is easy to carry around.

Apart from the parameters leaf area and fresh weight, the relative growth of butterhead lettuce from the time of iodine application to harvest was calculated using FM data from different trials. This demonstrated a decreasing iodine concentration with increasing pre-harvest intervals and the accompanying decreasing leaf application area (cf. Figure 5.11 C). Thus, the relative growth stage indicates, to some extent, the degree of dilutory effects which can be expected following a foliar application. Hence, the fresh mass at the time of application and the relative growth until harvest may be very useful parameters for better forecasting the iodine content of sprayed vegetable cultivations and more data should be collected in further investigations to design a robust and reliable prediction model.

The comparison between washed and unwashed samples treated one day before harvest showed significant salt residues on the leaf surface only with the KIO_3 treatment (cf. Figure 5.10 B). Previous observations of the postharvest treatments of culinary herbs (cf. Figure 5.6) showed no significant differences between washed and unwashed samples. Therefore, it can be deduced that foliar iodine uptake is a time-dependent process probably taking place within minutes or hours for KI and within days for KIO_3. If excess KIO_3 salts adhere to the leaf surface for a few days, two scenarios could occur: Sudden precipitations or overhead sprinkler irrigation can cause run-off of salt residues onto the soil or in the direction of the stalk in the inner part of the head, depending on the leaf curvature and constitution. If no irrigation or rainfall occurs, excess KIO_3 salt will liquefy (under favorable conditions of > 93.8 % RH) and eventually be absorbed by the leaf showing by hindsight no significant differences in iodine accumulation to the KI treatment. Therefore, further investigations at shorter application intervals before harvest and with overhead sprinkler irrigation treatments should be undertaken to evaluate the different foliar permanence characteristics of KI and KIO_3.

- **Water application rates**

The application of iodine salts on butterhead lettuce was tested at varying water application rates (cf. Figure 5.12). An increased iodine accumulation could be ascertained with increasing concentrations of iodine in the spray solution, i.e. with decreasing water supply, especially pronounced in KI treatments. The slightly higher accumulation levels in the KI treatments may be explained by its characteristic properties, i.e. higher hygroscopicity and a smaller ionic size than KIO_3, when sprayed under favorable environmental conditions.

Apart from a few exceptions, most aerially applied nutritional elements follow Fick's first law: Solutes move from a low to a high concentration domain proportionally to the concentration gradient, i.e. the larger the concentration difference, the higher the flux rate and the degree of uptake by plants (RIEDERER AND FRIEDMANN 2006). Although not being an essential element for plants, iodine follows the same law and it can therefore be sprayed successfully under varying water application rates. Low water application rates imply less soil compaction, fuel consumption and equipment costs. High water application rates generate a higher RH and a lower temperature on the leaf surface and contribute to improved spreading and wetting of aerially applied solutions (KOCH 2007). These features may prove to be decisive advantages for vegetable growers that tend to perform tank mixing with several different agrochemicals at varying water requirements, compound compatibility permitting. In such cases, the adaptation of the iodine dose to lower or higher water application rates must be considered and is easy to implement, e.g. the reduction of the iodine quantity by approximately 30 % spraying at 300 L H_2O ha^{-1} (cf. Figure 5.12).

- **Application system**

The comparison of the application efficiency between the hand-held sprayer system and the common operational system with a mounted field sprayer showed efficiency differences of > 50 % in favor of the experimental application (cf. Figure 5.13). The comparison between KI and KIO_3 treatments showed accumulation variability in dependency of the application system. A possible explanation for this pattern is a diminished fresh head weight in the KIO_3/mounted field sprayer treatment and the consequential higher iodine concentration (cf. Figure 5.13 A). The yield decrease can in turn be interpreted by taking fluctuations of the soil properties into account, e.g. soil compaction, texture and/or fertilizer heterogeneities in the very large plot area (cf. 5.3). The reduced application efficiency of the operational system by more than 50 % can be explained by considering the mechanical configuration of a field sprayer system: The regular spacing of nozzles on the spray boom leads to a more even distribution and a consequential wetting of the soil surface between the lettuce heads. Apparently, the fixed spraying height and the constant pressure lead to a higher spray drift in the presence of wind guts. In contrast, the hand held spray system produces a more irregular spray pattern and a varying spray height. Although trying to spray in regular patterns, manual application may be subject to beneficial actions favoring a well-wetted lettuce head, e.g. stopping to spray when wind gust emerge and the avoidance of spraying the gaps between plants. In addition, the spray gun was observed to generate a fine mist probably consisting of a larger number of smaller droplets compared to the field sprayer system. SCHREIBER AND SCHÖNHERR (2009) postulated increasing penetration rates of solutions with decreasing droplet radii. This strategy, in turn, appeared to be limited by the factor drift loss.

However, the wide differences observed in this comparison must be taken into account when calculating iodine application rates for commercial use. Furthermore, increased experimental systems with spray booms and exchangeable spray nozzles should be implemented to achieve a better reproducibility and comparability to the commercial equipment.

5.5.5 Effects of iodine foliar sprays on the phenolic content of two multi-leaf lettuce varieties

The total phenolic content of Salanova® multi-leaf lettuce could only be slightly (6.2 - 8.8 %), although statistically significantly, enhanced by the application of iodine foliar sprays. The overall concentration of total phenolic compounds was almost twice as high in red multi-leaf lettuce which is consistent with results obtained by LIU ET AL. (2007) who made investigations on secondary compounds of several lettuce cultivars. The study conducted by BLASCO ET AL. (2008) on hydroponically cultivated butterhead lettuce showed, in contrast to the results in section 5.4.4, a much greater increase of total phenolic compounds compared to the control treatment. In fact, up to a five-fold higher concentration without biomass impairment was found at 40 µM I^--I in the nutrient solution. On the other hand, at the same iodine fertilization level, a concentration of > 900 mg I (kg DM)$^{-1}$, corresponding to ≈ 3600 µg (100 g FM)$^{-1}$ at a supposed 4 % DM fraction, was determined. Considering the lowest iodine accumulation level achieved in the same study (10 µM IO_3^--I treatment), no significant differences in the total phenolic concentration were computed, but already more than 2000 µg I (100 g FM)$^{-1}$ were detected. The latter value is two times the upper tolerable daily intake level; it exceeds the recommended daily intake (RDI) of an adult person by a factor of 13 and thus cannot be recommended as biofortified butterhead lettuce for the daily diet. In our experiment on multi-leaf lettuce the iodine accumulation was only moderately above the RDI value (cf. Figure 5.14). On the other hand, significances on total phenolic compounds were also achieved on account of the little data deviation determined among the blocks repetitions. However, although the differences in total phenolic compounds were only minimal, it seems that iodine foliar sprays may positively influence the antioxidant capacity and therefore improve the nutritional quality of lettuce plants. Hence, further investigations into the impact of iodine doses relevant to biofortification practice on the antioxidant compounds of different lettuce varieties should be conducted to better understand and possibly validate the present results. In addition to the total phenolic compounds, consideration should be given to the determination of other important secondary metabolites such as total flavonoids, anthocyanins or vitamins, and also to different antioxidant capacity assays like FRAP, TEAC or DPPH (BENZIE AND STRAIN 1996; LIU ET AL. 2007; IGNAT ET AL. 2011). The remarkable differences in the iodine accumulation found between the green and red varieties cannot be justified by dilutory effects since the green variety showed less relative growth after the iodine application. The better permeability to iodine of the red variety may be explained by a different cuticular structure and composition of cuticular waxes or differences in stomatal size and density (cf. Figure 5.18).

5.5.6 Phloem mobility of iodine and its relevance for iodine accumulation in the edible parts of vegetables

The retranslocation of iodine in plants appears to be very limited (HERRET ET AL. 1962; BLASCO ET AL. 2008; VOOGT ET AL. 2010). This could be confirmed by our own foliar application trials conducted on kohlrabi where only low levels of iodine could be found in edible plant parts (cf. 4.4.4). Further foliar application trials on crisp lettuce, broccoli and white cabbage corroborated the innately poor translocation of iodine in several vegetable species (cf. 5.4.1). Nevertheless, the analysis of separated plant parts showed a reasonable iodine translocation from the leaf blade to the leaf stalks of kohlrabi (cf. 5.4.5). In case of butterhead lettuce (washed samples; cf. 3.3.1) a displacement of iodine into the stalks has been observed.

However, it is unclear if the translocation derives from the phloem mobility, from a preferential flow over the outer surface of the leaves at a later point in time during cultivation, or from the usage of both pathways. The latter results are in agreement with the observations of STRZETELSKY ET AL. (2010) and SMOLEŃ (2011b), who sprayed radish and carrots at doses of (2x) 0.8 kg I ha^{-1} and (4x) 0.005 - 0.5 kg IO_3^--I ha^{-1}, respectively. A rather improbable explanation for these observations is that the remaining iodine solution may have dripped onto the soil underneath and then followed a preferential flow-path over soil cracks. The most probable scenario though, is that the multiple iodine applications over a period of approximately 2 months allowed a slow, but consistent iodine translocation through the phloem. It is thus conceivable, that iodine can display low or high mobility depending on the vegetable species in a comparable manner to boron or, like the elements Fe, Zn, Cu and Mo, a conditional phloem mobility (BROWN AND HU 1996, 1998; BROWN AND SHELP 1997; EPSTEIN AND BLOOM 2005; MARSCHNER 2012).

5.5.7 Iodine uptake of foliage as affected by leaf topography

In preliminary investigations, the variations in stomata size and density as well as other leaf characteristics of different vegetable species were emphasized by observations with microscopical methods (cf. Figures 5.16 - 5.20).

FERNÁNDEZ ET AL. (2013) postulated that specific morphological and structural components of plants, like stomata or trichomes, are important uptake pathways for solutions on the leaf surface. Hence, the wide differences in stomata size and density encountered between species (winter spinach, butterhead lettuce and white cabbage) and between varieties within a single species (green and red multi-leaf lettuce) may partially provide consistent explanations for the distinctive iodine accumulation behavior observed. Winter spinach was the vegetable species with the largest stomata size and density and was found to accumulate iodine to the greatest extent (cf. Figure 5.16 and Table 5.5). Concordantly, white cabbage showed the lowest iodine content level and the smallest stomata size but comparable stomata density to winter spinach. Butterhead lettuce presented moderately large stomata at a lower density than winter spinach and takes an intermediate position between the previous two species. The comparison between the green and red multi-leaf lettuce further corroborates the importance of the stomata as a solute uptake pathway: The red variety has a stomata size almost double that of the green variety and a correspondingly twofold higher iodine accumulation in the iodate treatment, even with only half of the stomatal density (cf. Figure 5.14 B and Figure 5.18 A and B). The stomata density thus seems to be subordinate to the stomata size. In addition, the larger molecular size of IO_3^- may have substantial disadvantages compared to I^- when sprayed on the leaves of the green variety with a smaller stomata size compared to the red variety (cf. iodide and iodate treatments in Figure 5.14 B).

Therefore, vegetable species with a large stomata size and density may be prone to accumulate larger amounts of iodine and, in turn, species with a small stomata size and a low density may only accumulate small amounts of the same dose of aerially applied iodine (and most probably many other mineral nutrients). In addition, the possible combinations in between, i.e. a large density of small stomata and vice versa, may show large differences in their accumulation behavior, even exhibiting the same total stomatal pore area. However, even though the different stomatal sizes and densities cannot be assumed to be the sole components of the solute uptake system on leaves, they may be fairly good accumulation indicators for iodine and other mineral nutrients.

The structural comparison of different trichomes (cf. Figures 5.19 and 5.20) may further complicate the uptake model of solutes on leaves. Firstly, the total leaf absorption area will vary to a greater or lesser extent, depending on the form, size and density of the trichomes. Secondly, some trichomes shapes, particularly the shallow, glandular trichomes, may provide further retention and absorption pathways for the solutes (cf. Figure 5.20 B) through the cuticular depression molded during their formation.

This, together with the observations on the stomata, suggests a graduation in the solute absorption efficiency between different leaf surface structures. Hence, large stomatal pores may be more effective uptake pathways than small ones or trichomes. Glandular trichomes may be of greater importance for the solute uptake than hair-like ones and so on. However, although the wide genotypic differences of vegetable species make the quantitative characterization of the leaf surface topography rather difficult, further investigations should be conducted to assess these important uptake pathways. This represents a great opportunity to develop a reliable prediction model for the application of foliar sprays on crops.

6 General discussion

A better understanding of the effects of iodine applied to vegetable crops is of great importance for the knowledge on how to use the specific plant response to the advantage of growers and consumers. In this respect, the aim of this thesis was to develop and assess feasible iodine biofortification strategies for vegetable crops and estimate their specific fertilization level.

To attain this objective, several subgoals were processed and evaluated stepwise. These subgoals will be discussed in the following:

- Identify suitable **vegetable species** for iodine biofortification and their specific fertilization levels (6.1).
- Determine **environmental and cultivation factors** affecting iodine accumulation (6.2).
- Assess the most appropriate **iodine fertilization form** (6.3).
- Investigate the efficiency of **iodine application techniques** (6.4).
- Test a **fertilizer prototype and the miscibility of different agrochemicals with iodine** under practical conditions (6.5).

To better understand the complex relationship between the vegetable species, the environmental and cultivation factors as well as the influence of the solution formulation and the application technique on the iodine accumulation of vegetable crops, a schematic overview of important influencing factors is given in Figure 6.1 using the example of foliar sprays (further explanations are given in the subsequent sections).

Finally, **economic and marketing aspects** of iodine biofortification (6.6) will be elucidated and **suggestions for the improvement of the experimental methods** (6.7) will be given.

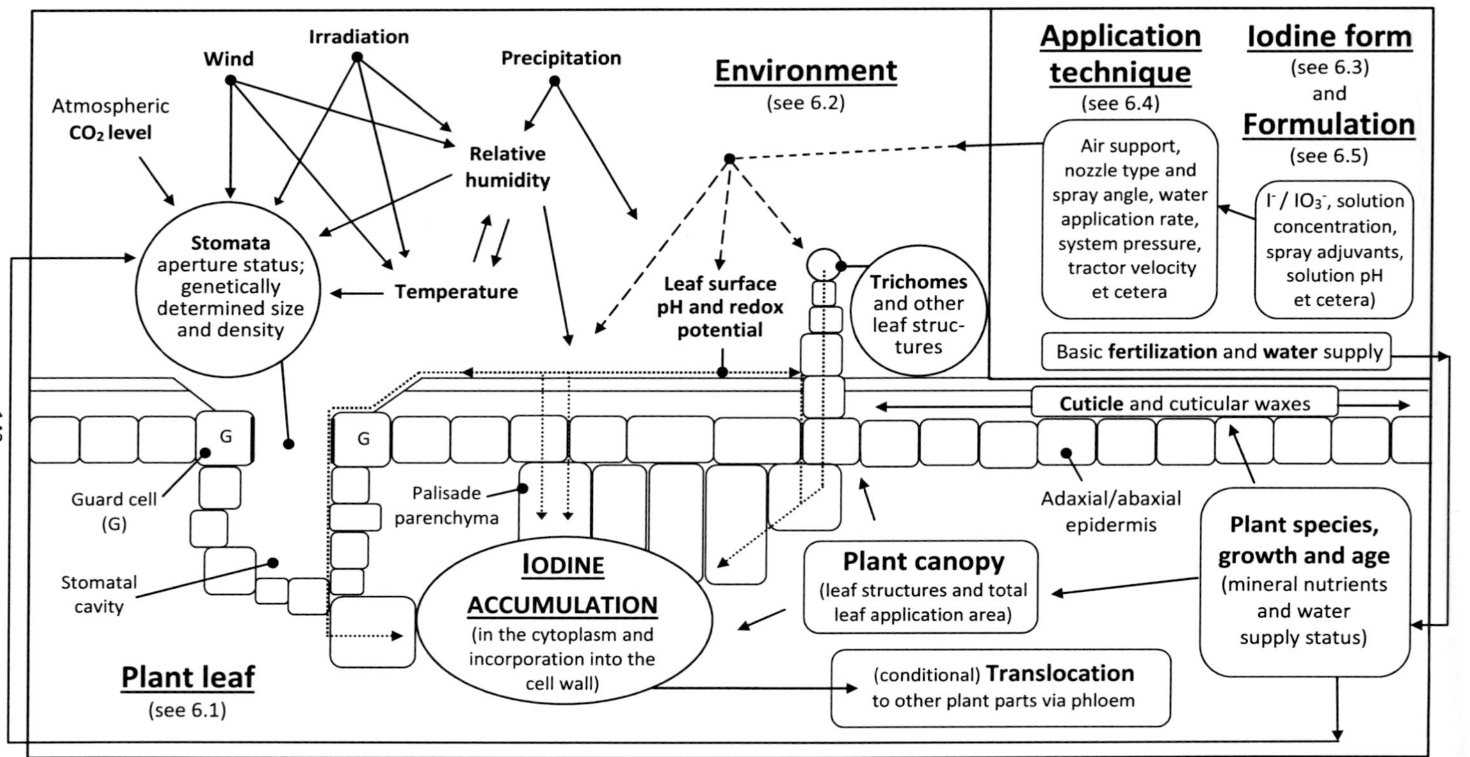

Figure 6.1 Simplified schematic representation of important influencing factors and accumulation pathways of iodine in plant leaves (see text for further explanations). Dashed arrows symbolize foliar sprays. Dotted arrows symbolize the diffusional movement of iodine solutes onto the leaf and possible apoplastic and/or symplastic uptake pathways

6.1 Suitable vegetable species for iodine biofortification

The identification of suitable vegetable species for iodine biofortification was strongly related to the iodine form and dose as well as the application technique. In Table 6.1 an overview is given of the vegetable species investigated, with emphasis on their general adequacy for iodine biofortification, in relation to the aforementioned factors.

Many vegetable species will probably accumulate iodine in edible plant parts to some extent provided that enough iodine is supplied over a certain period of time. However, the main constraints for iodine biofortification, i.e. a negative impact on crop yield and marketable quality as well as the costs for the iodine fertilizer and its application, make the pool of suitable vegetables smaller.

Based on the results of this thesis, it can be stated that **leafy vegetables** with a **loose habit** and a **large leaf surface** as well as a **low fresh mass** are generally the most appropriate vegetable species for iodine biofortification because of the following fundamental reasons: Firstly, a **loose habit** and a **large leaf surface** offer both a large transpiration area and consequentially an augmented xylematic transport of mineral elements applied to the soil as well as a large application area for foliar sprays and the associated enhanced mineral elements absorption. Secondly, leaves that exhibit a **low fresh mass** will need less total iodine to achieve a certain iodine concentration per mass unit compared to plant parts that exhibit a large fresh mass. Thirdly, the leaves of **leafy vegetables** constitute the absorption/accumulation organs and edible parts at the same time. This is a decisive advantage compared to other crop species that have a clear separation between the edible plant parts and the absorption/accumulation organs, e.g. in stem tubers and fruit vegetables (cf. 5.5.6).

The genus *Lactuca,* above all the species *L. sativa* var. *capitata*, was intensely investigated in this dissertation and found to be excellently suited for the biofortification of iodine, especially by means of foliar sprays. Single or twofold aerial KI and KIO_3 applications at doses in the range of 0.125 - 0.5 kg I ha^{-1} were well tolerated and a satisfactory iodine accumulation was achieved under all tested conditions. The fundamental suitability of butterhead lettuce for iodine biofortification has been corroborated by other recent studies where soil and foliar applications (SMOLEŃ ET AL. 2011a, 2013) as well as iodine applications by means of nutrient solutions were positively investigated (BLASCO ET AL. 2008; VOOGT ET AL. 2010; LEYVA ET AL. 2011).

Table 6.1 Overview of the investigated vegetable species with emphasis on their general adequacy for biofortification programs in dependency of the application form (F = foliar spray; S = soil application), iodine form (I^- = iodide; IO_3^- = iodate) and iodine dose (kg I ha^{-1}; see text for further explanations). * = to achieve 50 - 100 (or more in the case of culinary herbs) µg I $(100 g FM)^{-1}$ in edible plant parts. † = not tested application dose, i.e. recommended on basis of the results of the tested doses

Vegetable group	Species	Vernacular name	Tested application technique	Suggested iodine form	Suggested iodine dose * [kg I ha^{-1}]	Adequacy for biofortification
Leafy vegetables	*Lactuca sativa* L. var. capitata	**Butterhead lettuce**	S / F	IO_3^- (S)/ I^-; IO_3^- (F)	5.0† (S) /≈ 0.25 (F)	Good (S) /excellent (F)
	Lactuca sativa L. var. capitata cv. `Archimedes RZ´	**Multi-leaf butterhead green**	F	I^-; IO_3^-	≈ 0.25	Excellent
	Lactuca sativa L. var. capitata cv. `Gaugin RZ´	**Multi-leaf butterhead red**	F	I^-; IO_3^-	≈ 0.25	Excellent
	Lactuca sativa L. var. crispa	**Crisp lettuce**	S / F	IO_3^- (S)/ I^- (F)	≥ 7.5 (S) / ? (F)	Fair (S) / ? (F)
	Spinacia oleracea L.	**Winter spinach**	F	I^-; IO_3^-	0.05 - 0.1†	Conditional
	Diplotaxis tenuifolia L.	**Wild rocket**	F	I^-; IO_3^-	0.5 - 1.0	Good
	Petroselinum crispum [Mill.] Nym.	**Parsley**	F	I^-; IO_3^-	0.1 - 0.5	Excellent
	Allium schoenoprasum L.	**Chives**	F	I^-; IO_3^-	0.1 - 0.5	Excellent
	Origanum vulgare L.	**Oregano**	F	I^-; IO_3^-	0.1 - 0.5	Excellent
	Ocimum basilicum L.	**Basil**	F	I^-; IO_3^-	0.1 - 0.5	Excellent
Cabbages	*Brassica oleracea* L. convar. *capitata* var. *alba*	**White cabbage**	F	-	-	Inadequate (F)
	B. oleracea L. convar. *acephala* var. *gongylodes*	**Kohlrabi**	S / F	IO_3^- (S) / ? (F)	≥ 7.5 (S) / ? (F)	Fair (S) / ? (F)
	B. oleracea L. convar. *botrytis* var. *italica*	**Broccoli**	F	-	-	Inadequate (F)
Root and tuber vegetables	*Daucus carota* L. ssp. *sativus*	**Carrot**	S	-	-	Inadequate (S)
	Raphanus sativus L. var. *sativus*	**Radish**	S	IO_3^-	≥ 7.5	Fair (S)
	Allium cepa L. agg. *cepa*	**Onion**	S	-	-	Inadequate (S)

Other cultivars of the species *L. sativa* var. *capitata* with a similar habit, e.g. green and red varieties of batavia lettuce, baby leaf, leaf lettuce and oak leaf lettuce, just to name a few, will probably show a comparable iodine accumulation behavior at similar iodine doses. The tested culinary herbs (Table 6.1) showed throughout an excellent iodine accumulation behavior even at the lowest application dose of 0.1 kg I ha^{-1}. Although the iodine accumulation levels in culinary herbs ranged from 50 - 600 µg I (100 g FM)$^{-1}$, higher iodine application doses of up to 0.5 kg I ha^{-1} are supposably uncritical, because the fresh mass amounts of culinary herbs used as food ingredients are usually portions of merely a few grams fresh herb weight per meal (cf. 5.5.3.). It is probable that other leafy herbs like cress (*Lepidium sativum* L.), wild garlic (*Allium ursinum* L.), coriander (*Coriandrum sativum* L.), lovage (*Levisticum officinale* W. D. J. Koch), salad burnet (*Sanguisorba minor* Scop.) or hyssop (*Hyssopus officinalis* L.) will show a similar good iodine accumulation behavior at iodine doses in the range of 0.1 - 0.5 kg I ha^{-1}.

Although some of the vegetable species investigated (Table 6.1) showed barely sufficient (radish) or insufficient (carrot and onion) iodine accumulation when fertilized by means of soil applications, it is not certain that these crops are generally inadequate for iodine biofortification. As recorded by SMOLEŃ ET AL. (2011b) and STRZETELSKY ET AL. (2010), multiple foliar sprays on carrot and radish are most probably the better application alternative for these species.

The varied crop morphology and the existing phenotypic differences in the plant canopy, **cuticular constitution, stomatal size and density** as well as other leaf structures like trichomes, may all influence the iodine uptake and accumulation into the plant leaves (Figure 6.1). Although the mechanisms of penetration are currently not fully understood, it is certain that plant surfaces are permeable to nutrient solutions (FERNÁNDEZ ET AL. 2013) and iodine apparently belongs to the mineral elements that can be absorbed by plant leaves. The iodine solutes may penetrate the plant surface either by passing the hydrophobic leaf cuticle, the trichomes or by entering the stomatal pores by means of mass flow and diffusion.

The **stomata** are considered to be very important sites of entry for aerially applied nutrient solutions (EICHERT AND BURKHARDT 2001; SCHLEGEL ET AL. 2006) and species, or cultivars within a single species, with a large stomata size and/or density (e.g. winter spinach or red multi leaf lettuce; cf. Figure 5.16 - 5.18) may be more prone to accumulation of larger amounts of iodine than species with small stomatal pores and number per unit area (cf. 5.5.7). The leaf impairment observed during the experiments conducted on winter spinach and crisp lettuce showed that iodine doses of ≥ 0.5 kg I ha^{-1} may be excessive for some genotypes and thus modified biofortification strategies at lower iodine doses should be tested.

The **cuticular constitution**, i.e. its thickness or amount and composition of cuticular waxes may substantially constrain the iodine accumulation as seen in some species of the *Brassicaceae* family (cf. 5.5.2). The experiments on white cabbage and broccoli showed that cabbage varieties with a large fresh mass may generally not be recommended for iodine biofortification by means of foliar sprays. Other cabbages with a similar habit, such as cauliflower (*Brassica oleracea* L. var. *botrytis*) and its different cultivars (cauliflower green and cauliflower romanesco), and different cultivars of cabbage (*Brassica oleracea* L. convar. *capitata*) such as red cabbage, savoy cabbage and pointed cabbage may also not be suitable for iodine biofortification. However, some other species of the genus *Brassica* may possibly be suitable, especially when the leaves constitute the edible plant parts and the habitus is rather loose as in the case of Asia greens. Thus, different varieties of leaf mustard (*Brassica juncea* L. Czern cv. `Agano´; cv. `Frizzy joe´) mizuna (*Brassica rapa* L. var. japonica cv. `Mandovi´; cv. `Arun´), pak choi (*Brassica rapa* subsp. *chinensis* L. Hanelt. cv. `Arax´; cv. `Sagami´) or tatsoi *(Brassica rapa* L. var. rosularis cv. `Tama´) may indeed be good candidates for iodine biofortification among the *Brassicaceae* family.

Once iodine is absorbed and taken up by the leaf cells (Figure 6.1) it might be stored in the cytoplasm, incorporated in the cell wall (WENG ET AL. 2013) or translocated via the phloem to other plant parts (LANDINI ET AL. 2011; CAFFANGNI ET AL. 2012). Experiments on the phloem mobility of iodine in kohlrabi showed a limited but consistent iodine flux through the phloem (cf. 5.5.6). Hence, the translocation of iodine in vegetable crops may generally be assessed as being moderate and the used biofortification strategy, i.e. the **iodine dose and concentration** as well as the **number of applications** and the **time between** each application will substantially influence the overall translocation of iodine through the phloem and its accumulation into edible plant parts (STRZETELSKY ET AL. 2010; SMOLEŃ ET AL. 2011b).

6.2 Environmental and cultivation factors affecting iodine biofortification

The soil application experiments in this thesis showed that several vegetables cultivated after a one-time iodine soil fertilization at rates of ≤ 7.5 kg IO_3^--I ha^{-1} accumulated iodine in edible plant parts to a satisfactory amount [≥ 50 µg I $(100\ g\ FM)^{-1}$]. However, vegetables grown on the same plots in the second season without further iodine biofortification did not accumulate iodine to an adequate extent (cf. 4.5.2). Hence, long term effects of iodine biofortification by means of soil drenches could not be observed.

The results obtained in this thesis demonstrate that iodine added to arable soil is rapidly converted from plant available to plant unavailable iodine forms. The main factors affecting the iodine uptake of crops from soil are the processes of immobilization, adsorption, desorption, dissolution, volatilization and leaching. These processes are dependent on the soil type and its bio-physico-chemical properties, namely the organic matter content, redox potential, pH value, soil texture, iron and aluminum oxides, microbial and enzyme activity, clay minerals and water-logging (JOHANSON 2000; HOU ET AL. 2003; JOHNSON 2003; FUGE 2005; YAMAGUCHI ET AL. 2006; AMACHI 2008; ASHWORTH 2009; TANAKA ET AL. 2012). Several studies reported that the main sink of iodine in soils is apparently its incorporation into organic matter and that temperature heavily affects the incorporation rate. Iodide is thereby lost more rapidly, probably within minutes to hours and iodate within hours to days after its exogenous application to soil (WHITEHEAD 1973; SHEPPARD AND THIBAULT 1992; BOSTOCK ET AL. 2003; STEINBERG ET AL. 2008; SHETAYA 2011).

Therefore, a specific and precise iodine fertilizer placement based on the cultivated crop and its rooting depth is recommended. It is of major importance to provide vegetable crops with iodine in the most appropriate cultivation period, i.e. not before planting or sowing, but preferably at a middle or more advanced stage of the crop development. Hence, either a just-in-time fertilization or the use of seaweed as a slow-release iodine fertilizer may result in a more effective iodine biofortification strategy (HONG ET AL. 2008; WENG ET AL. 2008a).

The application of iodine by means of foliar sprays was intensely investigated in field and greenhouse experiments and this approach was ascertained to be a powerful and effective method of enhancing the iodine levels in several vegetable crops. A satisfactorily iodine accumulation in edible plant parts was obtained at relatively low doses of 0.1 - 0.25 kg I ha^{-1} and the superior performance of aerially applied iodine salts in comparison to soil applications will probably make foliar sprays prevail in future biofortification programs. However, wide differences in the efficacy of this approach were observed under fluctuating environmental conditions and heavily influencing factors such as light radiation, relative humidity and temperature at the application time could be individuated (cf. Table 5.5 and Figure 5.9). Accurately predicting iodine accumulation in vegetable crops is therefore difficult.

Even though few references are available about the main environmental and cultivation factors affecting iodine absorption and uptake by above-ground plant parts, it can be noted that **basic fertilization, water supply** as well as **climatic conditions** may consistently affect the aerial application of iodine (Figure 6.1). The availability of plant nutrients and water in the root zone will influence the nutritional and water status of a crop, and consequently its growth, development, cuticular constitution and the stomata aperture status; which all influence the accumulation of mineral elements applied on the crop foliage (FERNÁNDEZ ET AL. 2013). Once the iodine solution is applied to the crop leaves, a variety of environmental factors directly or indirectly affect the iodine absorption and uptake. The current **wind, solar radiation** or **precipitation** at or after the application have a direct impact on the RH and the temperature surrounding the plant leaves. All these climatic factors may influence, along with the atmospheric **CO_2-level**, the stomatal aperture status and thus the iodine uptake. **Precipitation** on crops directly after the application of foliar sprays may dilute the solution and cause a certain degree of run off of solutes from the leaves depending on the formulation of the solution (MARSCHNER 2012). The **RH and temperature** at the time of application majorly influence the drying process of the spray solution (FERNÁNDEZ AND EICHERT 2009) and consequently the available uptake time for the iodine solutes. Therefore, further investigations under controlled climatic conditions are needed to better understand the influence of the above mentioned factors and to possibly better forecast the iodine accumulation in vegetable crops (see below in 6.7).

6.3 Uptake of different iodine forms by roots and above-ground plant parts

Apart from the iodine dose, the response of vegetables to the iodine application by means of soil fertilization was determined primarily by the applied iodine form (I^-/IO_3^-). Satisfactory iodine accumulation levels in butterhead lettuce, crisp lettuce, kohlrabi and radish were mainly found using KIO_3 as the iodine fertilizer (cf. 4.5.2 and 4.5.3). Hence, the most recommendable iodine form for soil applications is **iodate**. These findings are consistent with the results of many other studies where a distinctively higher iodine accumulation was recorded when different crops were fertilized with iodate as compared to iodide (WHITEHEAD 1975; MURAMATSU ET AL. 1995; MACKOWIAK AND GROSSL 1999; ZHU ET AL. 2003; DAI ET AL. 2006; WENG ET AL. 2014). Even if plant roots absorb I^- at a higher rate than IO_3^- it seems that iodide is more subject to cumulative losses in the soil environment than iodate. The slower uptake rate of IO_3^-, which was often observed in nutrient solution experiments, is probably limited by the reduction process as well as the heavier molecular weight and the higher valence. However, the longer residence of IO_3^- in soil compared to I^- makes the oxidized form more phytoavailable which results, under field conditions, in higher iodine accumulation levels in edible plant parts (BÖSZÖRMÈNYI AND CSEH 1960; WHITEHEAD 1973, 1975; MACKOWIAK AND GROSSL 1999; ZHU ET AL. 2003; BLASCO ET AL. 2008; WENG ET AL. 2008a; VOOGT ET AL. 2010).

The intensely investigated iodine application by means of foliar sprays demonstrated a comparable foliar uptake between KI and KIO_3 treatments under controlled environmental conditions in greenhouses (cf. 5.4.2). However, field trials using butterhead lettuce as a model system indicated, under certain circumstances, a higher effectiveness of potassium iodide (cf. Figure 5.9) that was mainly explained by its higher hygroscopicity, the lower point of deliquescence and a smaller ionic size in comparison to potassium iodate (cf. 5.5.4). STRZETELSKY ET AL. 2010 recorded leaf and yield impairment on radish only using KI as foliar fertilizer (KI and KIO_3 were compared). Although in this study no further information is given about the iodine concentration obtained in edible plant parts, it can be assumed that the observed phytotoxic effects are probably due to a higher iodine accumulation rate in the KI treatment.

Figure 6.2 displays the results of data collected from 10 different trials on butterhead lettuce during the experimental seasons 2010, 2011 and 2012. The relation between the iodine

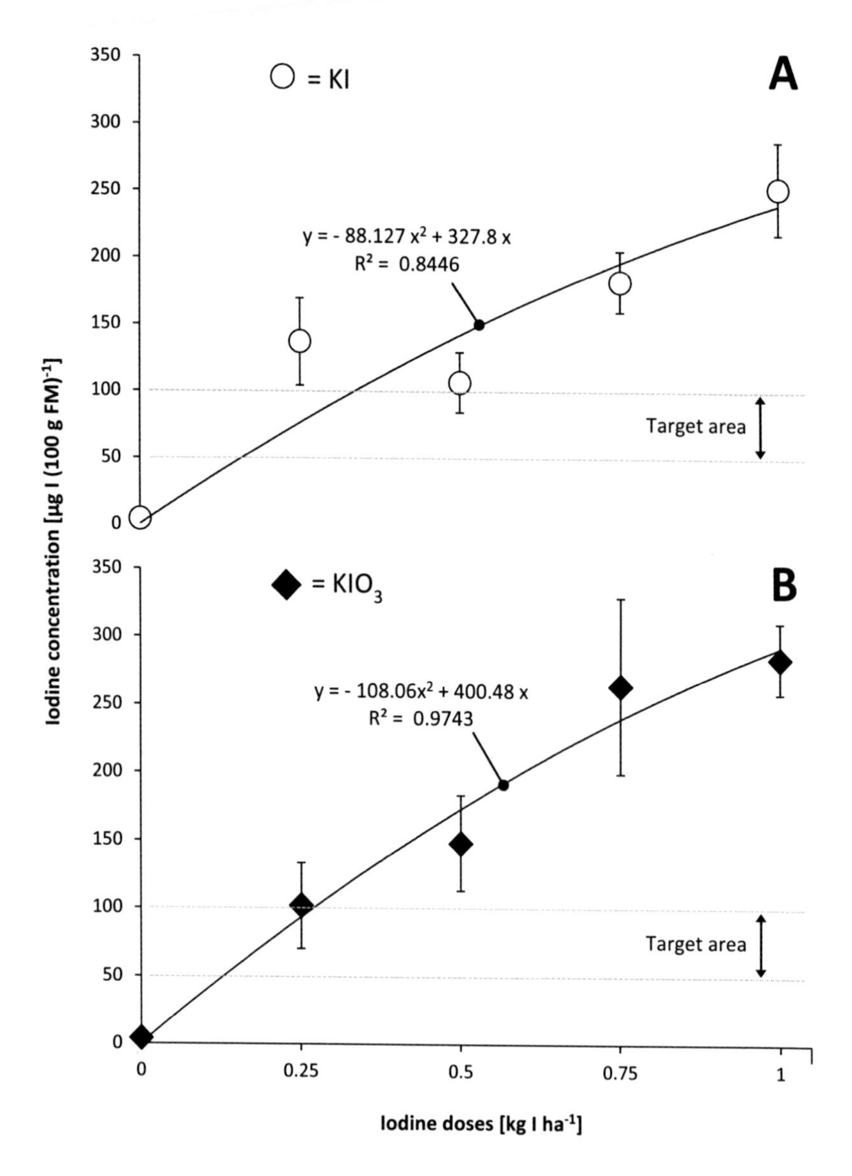

Figure 6.2 Relation between the iodine concentration in edible plant parts of butter-head lettuce and different iodine doses applied by means of foliar sprays. A = KI treatments = empty circles. B = KIO_3 treatments = diamonds. Data collected from 10 different trials during the experimental seasons 2010, 2011 and 2012

concentration in the edible plant parts and the different iodine doses applied by means of foliar sprays shows a similar correlation curve for both iodine forms. Therefore, both **iodide and iodate** come into question when applied by means of foliar sprays. However, the best coefficient of determination was calculated for the iodate treatments (Figure 6.2 B) which suggests a more reliable iodine accumulation predictability using potassium iodate as the iodine fertilizer. In addition, iodate is more stable and has a longer shelf-life in comparison to KI (PAHUJA ET AL. 1993; DIOSADY ET AL. 2002). The advantages of iodide are a lower POD, higher solubility (GREENSPAN 1977; APELBLAT AND KORIN 1998) and a partially faster uptake by plant leaves (cf. 5.5.4). Hence, both iodine forms have some advantages that may be combined. The iodine accumulation behavior of leafy vegetables after the application of I^- and IO_3^- salts mixed at varied ratios should therefore be investigated in further trials.

6.4 Comparison of application techniques for iodine biofortification

The **soil application** of iodine salts has been thoroughly investigated in the last decade and the results have always shown a general adequacy of the pathway soil for the iodine biofortification of vegetables (DAI ET AL. 2004a, 2004b; HONG ET AL. 2008; WENG ET AL. 2014).

The results in this dissertation revealed that the iodine quantities needed to achieve a satisfactory accumulation in edible plant parts of butterhead lettuce were fifteen to thirty times higher compared to the foliar applications (7.5 kg I ha^{-1} compared to 0.25 - 0.5 kg I ha^{-1}). In addition, long term effects of a one-time iodine fertilization were not recorded (cf. 4.5.2). However, as observed on kohlrabi and crisp lettuce, the comparison between foliar and soil applications (cf. 4.5.5, Figure 4.7 and Table 5.5) showed a better adequacy of soil applications to obtain the desired iodine accumulation in edible plant parts. Hence, either a modified foliar application strategy with multiple sprays (cf. 5.5.6) or an optimized soil fertilization strategy should be implemented and tested in future investigations (as discussed in section 6.2). Apart from that, other fields of application for iodine in horticulture, such as the use as an herbicide, plant booster or iodine as a beneficial element against salinity stress have been reported (MYNETT AND WAIN 1971, 1973; SLIESARAVIČIUS ET AL. 2006; LEYVA ET AL. 2009; SZWONEK 2009). Hence, the soil application of iodine may produce additional advantages in the cultivation of crops and these possibilities should be therefore considered and tested in future trials.

The use of **foliar sprays** for the iodine biofortification of leafy vegetables was found to be more effective, easier to apply and economically more justifiable compared to the iodine soil applications. However, the desired iodine concentration in edible plant parts in the range of 50 - 100 µg I $(100\ g\ FM)^{-1}$ was not easy to achieve because of the multitude of different factors influencing the aerial application of iodine on vegetable crops (Figure 6.1). To date, no studies are available on the influence of commercial application equipment on the efficiency of iodine foliar sprays. Nevertheless, the comparison of the application efficiency between the hand-held sprayer system and the common operational system with a mounted field sprayer showed efficiency differences of > 50 % in favor of the experimental, hand-held applicator (cf. Figure 5.13). The reduced application efficiency of the operational system may be explained by considering the regular spacing of nozzles on the spray boom and the consequential wetting of the soil surface between the lettuce heads. Furthermore, the fixed spraying height and the constant system pressure may lead to a higher spray drift in the presence of wind guts. In addition, the experimental spray gun was observed to generate a fine mist probably consisting of a larger number of smaller droplets compared to the field sprayer system. Decreasing droplet radii may result in increasing penetration rates of foliar spray solutes (SCHREIBER AND SCHÖNHERR 2009). Hence, the efficiency of iodine foliar sprays may vary widely depending on the used system. Several different commercial spray systems are currently available: With or without additional air support, electrostatic spray systems or systems with "dropleg" devices. The many different interchangeable nozzles types will produce a range of droplet radii varying from very fine to very coarse (10 - 1000 µm). In addition, there are a multitude of applicable spray angles and patterns. The water application rate, the hydraulic system pressure and the velocity of the tractor; all these mechanical factors may consistently influence the results of the foliar application (VAN DE ZANDE ET AL. 2003; RUEEGG ET AL. 2006; ARBUCKLE 2012). Therefore, it is of major importance to investigate the influence of different commercial spray systems on the efficacy of iodine sprays.

6.5 Iodine fertilizer prototype and miscibility of different agrochemicals with iodine

The miscibility trials conducted on butterhead lettuce with pure KI and KIO_3 salts as well as with the iodine fertilizer prototype showed that the formulation of the foliar sprays applied may affect the iodine accumulation in edible plant parts. The defining factors being the iodine form and dose, the solution concentration as well as the mode of action and the concentration of the added agrochemicals (cf. 5.5.4).

Tank-mixing different agrochemicals (pesticides, fertilizers) is labor-saving and thus a common agricultural practice which may produce economic advantages for the grower (GRIFFITH 2010). In addition, soil compaction caused by agricultural machineries can be further reduced. In order to improve the efficiency of foliar sprays, many spray adjuvants are available on the market with different modes of action: Surfactants lower the surface tension, stickers increase the solution retention and rain-fastness, neutralizers buffer the solution pH and acidifiers lower the solution pH. Penetrators increase the rate of foliar penetration by solubilizing cuticular components, compatibility agents improve the compatibility between different compounds, humectants retard the drying of the solution on the leaf by lowering the POD and drift retardants allow better spray targeting and deposition on foliage. Hence, the solution pH and other chemical parameters of the many possible solution formulations can differ widely and may consequently affect the physico-chemical properties of plant cuticles and finally the sorption of exogenously applied ions (SCHÖNHERR AND HUBER 1977; SCHÖNHERR 2000; FERNANDEZ AND EICHERT 2009; FERNÁNDEZ ET AL. 2013).

The iodine fertilizer prototype formulated as an emulsifiable concentrate was tested under field conditions and showed a comparable performance to the corresponding dose of pure KIO_3 salt and spray adjuvant and can therefore be recommended for commercialization. The miscibility experiments of the IFP and the pure iodine salts with the commercial products Calcinit®, Stopit®, Karate® and Revus® resulted in a good compatibility of the compounds and, in certain cases, in an improved iodine accumulation (cf. 5.5.4).

Hence, if the miscibility of compounds permits blending, it would certainly be meaningful to combine other mineral elements often lacking in the human diet in a single foliar spray solution in order that iron, zinc, copper, calcium, magnesium or selenium (WHITE AND BROADLEY 2009) salts might be blended with iodine and possibly with different pesticides.

However, currently only little is known about the multiple interactions of iodine with many different mineral elements, pesticides and spray adjuvants and thus it would be reasonable to continue miscibility trials with iodine. The physico-chemical compatibility of compounds may be tested in advance by carrying out the so-called "jar test" according to the American Society for Testing and Materials E 1518 - 05 method (ASTM 2012). Instable mixtures will eventually settle out, flocculate, foam excessively or poorly disperse. This may reduce the foliar spray efficiency and/or cause the clogging of sprayer nozzles and screens. Once the physico-chemical compatibility is positively tested, the biological compatibility should be tested on a few target plants. A biological incompatibility will produce a loss of effectiveness of one or more compounds and the possible interactions between several chemicals may have a phytotoxic impact.

6.6 Economic and marketing aspects of iodine biofortification

The economic viability can be considered as crucial for the implementation of a new agricultural technique and a notional calculation based on the cultivation of butterhead lettuce will therefore be exemplified in following: Considering that the raw material price of approximately 41 US$ (kg I)$^{-1}$ (POLYAK 2013) and an exchange rate of ≈ 1.35 UD$ €$^{-1}$, the total cost for an iodine soil application at a fertilizer rate of 7.5 kg IO_3^--I ha^{-1} would amount to approximately 230 € ha^{-1}. Appling foliar sprays at 0.25 kg I ha^{-1} the iodine fertilizer costs would decrease to ≈ 8 € ha^{-1} and additional equipment and personnel costs for the grower need not arise if the product is tank-mixed with other agrochemicals. A cost calculation for the application of iodine in relation to the total costs incurred for the cultivation of butterhead lettuce is displayed in Figure 6.3. Hence, an additional iodine fertilization applied by means of foliar sprays (Figure 6.3 A) would amount to 0.07 % of the total costs (= 5.4 % of the fertilizer costs) whereas the soil application (Figure 6.3 B) would incur with a share of 2.58 % of the total costs (= 62.4 % of the fertilizer costs). Considering a crop density of 70,000 lettuce heads ha^{-1}, an additional 0.01 € - cent head^{-1} (foliar sprays) or 0.3 € - cent head^{-1} (soil applica-

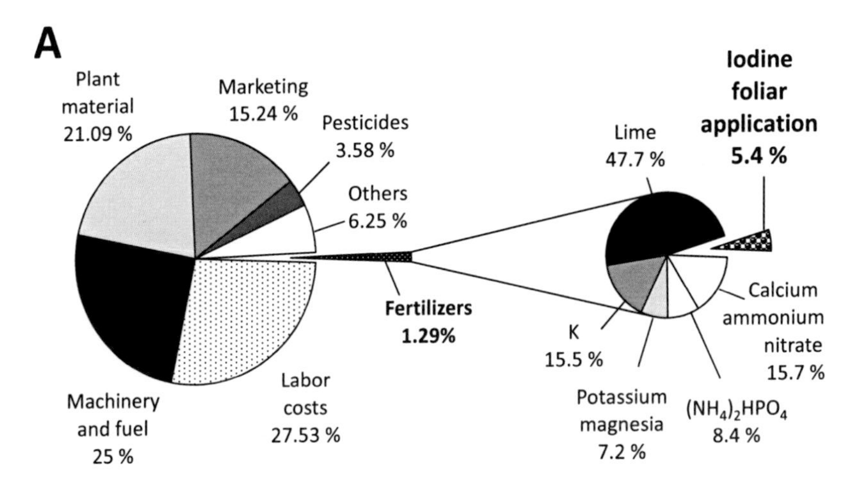

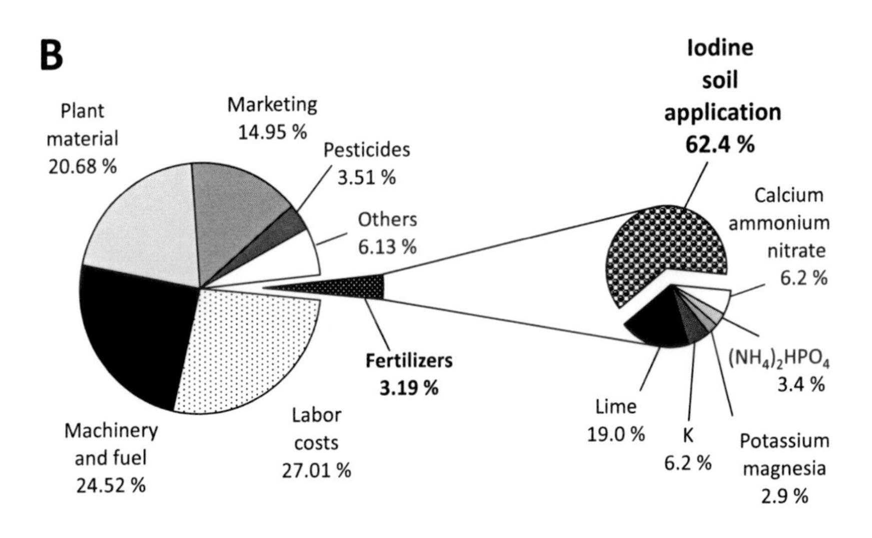

Figure 6.3 Share of the total costs for different iodine applications. A = foliar sprays. B = soil application. Mixed calculation based on the successive planting of butterhead lettuce (2x early planting under agricultural foil and 8x midseason planting). Cropping area = 5 ha; average distance between fields and farm = 5 km (KTBL 2009)

tion) would be necessary to offset the costs incurred and are therefore negligible, especially if the benefits of iodine can be advertised on the package of the vegetable products which may in turn produce a marketing advantage over the competitors.

The Italian company Pizzoli S.p.A. is the patent holder of biofortified potatoes and started the production and sale in the year 2008 (Zanirato 2008). The product range has now been extended to include tomatoes, carrots and (hydroponically grown) salad mixes (Figure 6.4 B). According to the EU Regulation no 1169/2011 (EU 2011) on the provision of food information to consumers, additional declarations for nutrients can be made, if these are present in significant quantities, next to the mandatory (will become effective on the 01 December 2014) and facultative nutrition declarations (Figure 6.4 A). Nutrient amounts in "significant quantities" are achieved when 15 % of the recommended allowance for a nutrient per 100 g or 100 mL is reached. With the recommended daily allowance for iodine amounting to 150 µg, the minimum needed iodine concentration is 22.5 µg $(100\ g\ FM)^{-1}$ which can be easily reached using the biofortification strategies investigated in this thesis.

In addition, the EU Regulation no 1924/2006 (EU 2006) on nutrition health claims made on foods states that claim such as "iodine-rich" on the food packaging are allowed if the concentration of the declared nutrient is twice as high as the significant quantity needed according to the EU Regulation no 1169/2011. Hence, the claim "iodine-rich" on packages is possible when the food contains iodine in concentrations amounting to 45 µg I $(100\ g\ FM)^{-1}$ or more. If the product was naturally enriched, i.e. biofortified with iodine, it can be additionally declared as "naturally rich in iodine". According to the list of permitted health claims made on foods in the EU Regulation no 432/2012 (EU 2012), iodine-rich food can be claimed in the following ways:

- Iodine contributes to normal cognitive function
- Iodine contributes to normal energy-yielding metabolism
- Iodine contributes to normal functioning of the nervous system
- Iodine contributes to the maintenance of normal skin
- Iodine contributes to the normal production of thyroid hormones and normal thyroid function

Therefore, these possibilities of advertising functional food should be exploited in order to enhance the economic impact of biofortified vegetables.

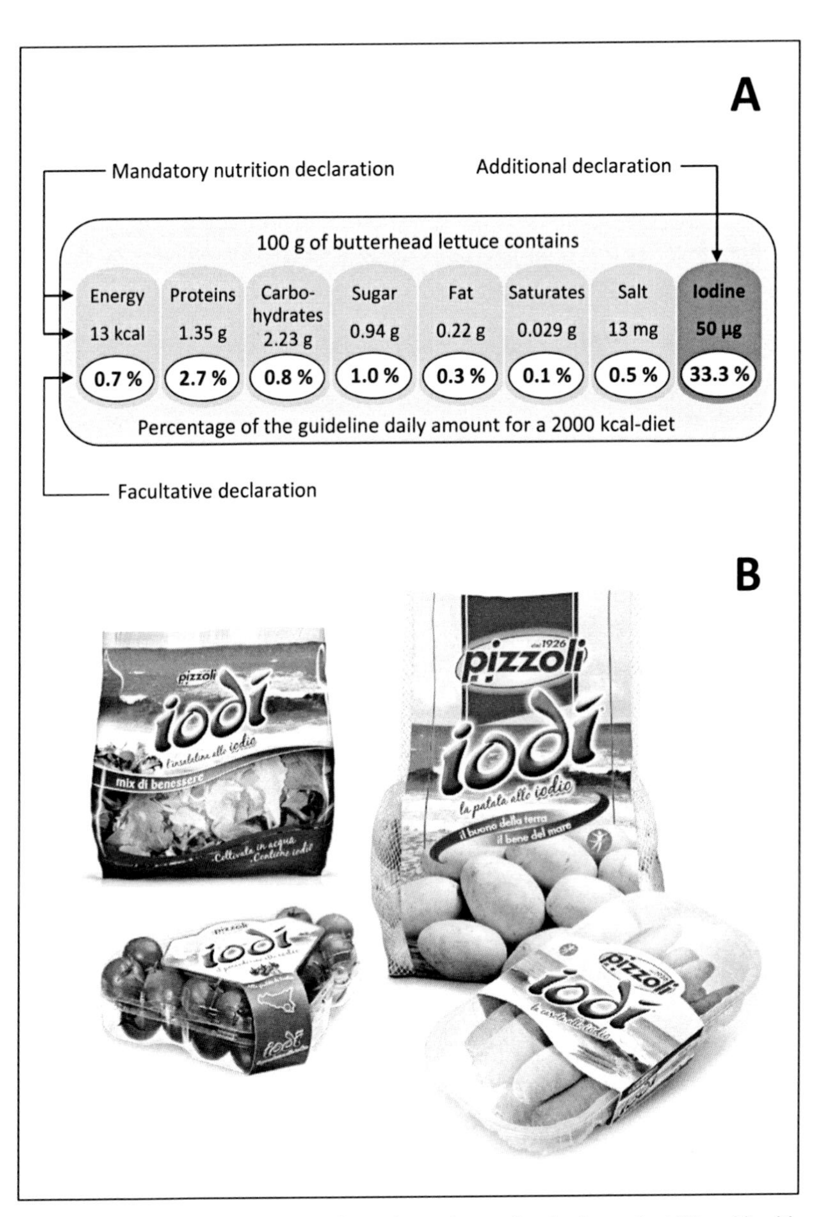

Figure 6.4 A = examples for the labelling of mandatory, facultative and additional health claims on packed goods according to EU (2006, 2011, 2012). Nutritional values for butterhead lettuce according to USDA (2011). B = Iodí® product line; vegetables biofortified with iodine produced by Pizzoli S.p.A. (picture by GGP 2012)

6.7 Suggestions for the improvement of the experimental methods

The extensively tested application of iodine by means of foliar sprays showed that: Firstly, the results of the experimental application system used in this dissertation are not directly conferrable to common operational devices and must therefore be either arithmetically corrected (cf. 5.3 and 5.5.4) or the experimental application systems have to be further improved. Secondly, the results of the field trials showed that environmental factors substantially influence the performance of foliar sprays (cf. 6.2) and should be further investigated in more detail under controlled ambient conditions.

The experimental system for field trials could easily be improved by adapting a commercial small plot spray boom (Figure 6.5 A) as required: The boom could be cut to length (e.g. to 1.5 m plot width) and adapted with three nozzles at a spacing of 50 cm. It should be further equipped with a precise dosage device to guarantee the homogeneous application of a specified iodine amount per unit of area. For example by installing an interchangeable 1000 mL spray solution PE-bottle between the air-pressure gauge and the first nozzle (schematic representation in Figure 6.5 A).

The influence of environmental factors can be investigated most conveniently by cultivating crops in climate chambers where temperature, solar irradiation and RH can be precisely adjusted to suit the requirements of the experimental issue. Unfortunately, however, climate chambers are usually of limited size (1 - 10 m²) and experiments with crops like butterhead lettuce will result in small trials with only a few individuals. This fact makes a statistical evaluation difficult and an analysis of variance impossible. Hence, for experiments in climatic chambers, an improved small-scale experimental unit as shown in Figure 6.5 B might be very useful. Lamb´s lettuce (*Valerianella locusta* L.) would be a particularly suitable model crop because of the following reasons: First preliminary trials (unpublished results) showed an excellent iodine accumulation behavior, the space requirements are minimal, the germination period (7 to 21 days) and the following cultivation period (4 - 6 weeks) are rather short and it can be grown all year-round. Hence, it could be grown outside, in the glasshouse or in large climatic chambers using commercial plastic trays sizing 50 x 70 x 10 cm filled with peat substrate. The cultivation density and the spacing between the rows must correspond to typical

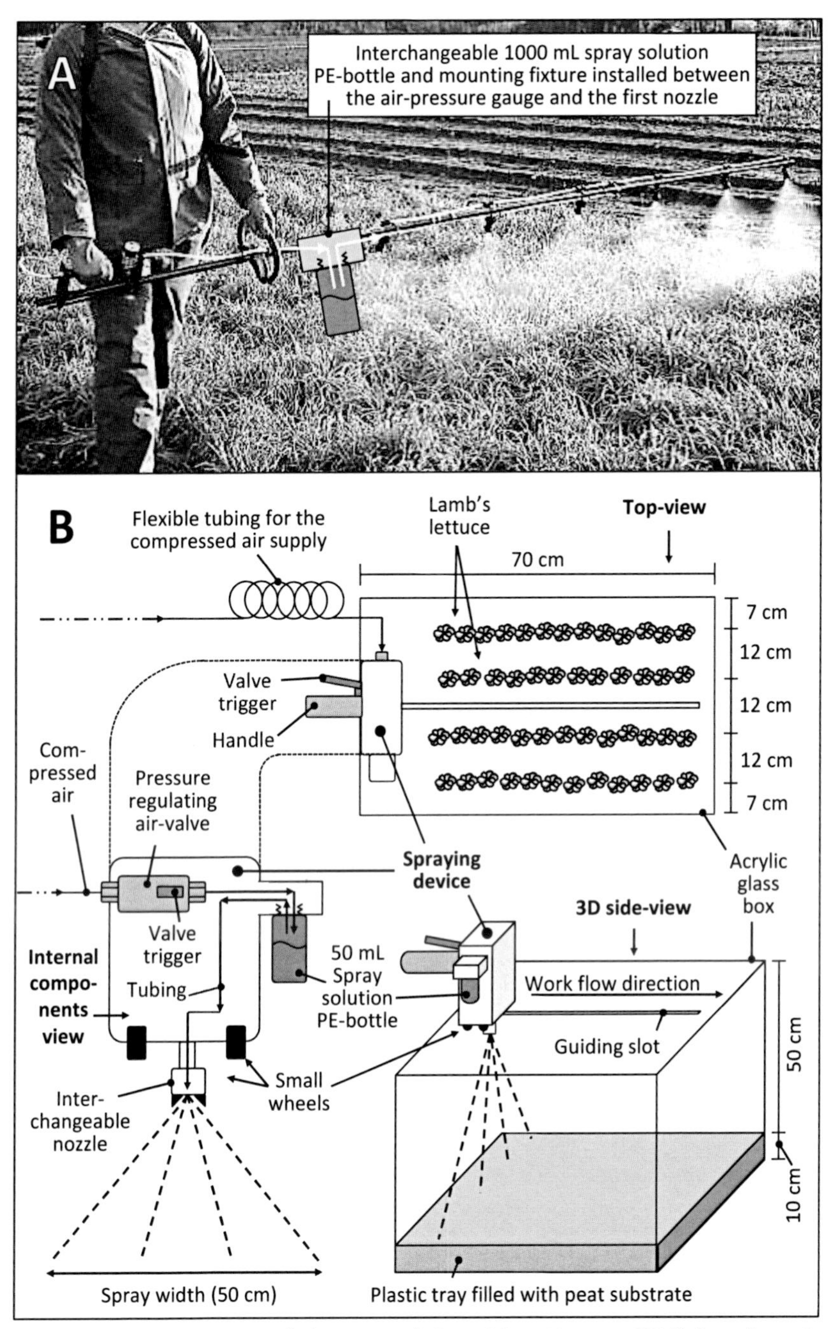

Figure 6.5 A = rigid boom small plot sprayer (adapted from HERBISEED 2009). B = schematic diagram (not to scale) of an experimental spraying device for greenhouses or climatic chambers

values used in the field (850 - 1250 kernel m²; 10 - 12 cm) or greenhouse cultivations (800 - 1200 kernel m²; 7 - 10 cm; WONNEBERGER ET AL. 2004).

The foliar spray solution could then be applied by means of the experimental spraying device under controlled environmental conditions in a climatic chamber. For this purpose, an acrylic glass box with a guiding slot will be positioned on top of the plastic tray to guarantee a reproducible and equally distanced spray application (Figure 6.5 B bottom right). The spaying device must be equipped with a handle and small wheels to be easily slid over the acrylic glass box. The main components of the spraying device are a pressure regulating air-valve with a trigger, a mounting fixture for the interchangeable spray solution bottle and a mounting fixture for the interchangeable nozzle.

The pressure regulating air-valve with a trigger allows for the connection of flexible tubing for the compressed air supply and for regulating the air-flow though the valve trigger into the internal tubing system. The different spray solutions needed for a dose-response trial can be easily interchanged by screwing treaded 50 mL PE-bottles into the mounting fixture for the spray solution bottle. Considering the cultivation area of the experimental unit is 0.35 m², a total spray solution volume of 35 mL (at 1000 L H_2O ha^{-1}) will be needed, but can be varied depending on the water application requirements (e.g. 21 mL at 600 L H_2O ha^{-1}, 10.5 mL at 300 L H_2O ha^{-1} and so on). Different commercial nozzle types could be mounted at the bottom of the spraying device. The mounting fixture of the nozzle will additionally serve as a guiding bolt in the guiding slot of the acrylic glass box. The foliar spray application must be carried out in a rapid but homogeneous work flow up to the stop of the guiding slot by grasping the handle and pressing the lever of the air-valve trigger. Before performing the first application, trial applications should be conducted on test surfaces, e.g. on blotting paper.

7 Summary

Iodine is an essential trace element for humans which is not ingested in sufficient quantities in many regions of the world. In spite of the wide use of iodized table salt, large portions of the German population are still inadequately supplied with the trace element. Therefore, further efforts are needed to improve the alimentary intake of iodine. The aim of this thesis was to develop an agronomical biofortification strategy for enriching vegetables with iodine by means of fertilization with iodine salts. Hence, foliar and soil application approaches were implemented and compared using a selection of different vegetable species in order to detect the influence of various environmental, application and soil factors affecting the iodine uptake by plants.

The iodine biofortification of vegetables by means of soil fertilization and foliar sprays with KI and KIO_3 salts was investigated in field and greenhouse experiments during the course of three experimental seasons (2010 - 2012) at different locations in Lower Saxony and North Rhine-Westphalia, Germany.

With rising iodine doses of KI or KIO_3 (0, 1.0, 2.5, 7.5 and 15 kg I ha^{-1}) applied directly to the soil before planting or sowing the iodine concentration in the edible plant parts of some vegetable crops increased. The highest iodine accumulation was observed in leafy vegetables (butterhead lettuce, crisp lettuce) followed by tuber vegetables (kohlrabi and radish). In these crops, the desired iodine content [≥ 50 µg I (100 g FM)$^{-1}$] was obtained or nearly achieved at a fertilizing rate of 7.5 kg IO_3^--I ha^{-1} without a significant yield reduction or degradation in the marketable quality. In contrast, the KI supply at the same rate resulted in much lower iodine enrichment and a yield reduction of > 10 % in some crops. Compared to the aforementioned vegetable species, root vegetables (carrots) and onions showed minimal iodine accumulation potential in their edible plant parts. After fertilizing KI or KIO_3, the iodine applied was rapidly converted to non-phytoavailable iodine forms; concordant with this finding, long-term effects of one-time iodine soil fertilization were not observable.

A comparison between the soil and the foliar fertilization techniques revealed a superior performance of aerially applied iodine on butter head lettuce. Therefore this biofortification approach was studied in detail on several vegetable species in further investigations. With increasing iodine supply (0, 0.1, 0.25, 0.5, 0.75, 1.0 and 2.0 kg I ha^{-1}) to above-ground plant parts, the iodine concentrations in edible plant parts increased when vegetables were sprayed with KI and KIO_3 on different dates close to the harvest. The highest iodine accumulation was observed in leafy vegetables (winter spinach, butterhead lettuce, rocket, basil, parsley and oregano) and the desired iodine content was already obtained at low fertilization rates of 0.1 - 0.25 kg I ha^{-1} without a significant yield reduction or degradation in the marketable quality.

Iodine applications under controlled environmental conditions in greenhouses demonstrated a comparable foliar uptake between KI and KIO_3 treatments. However, further field trials using butterhead lettuce as a model system showed greater efficacy of potassium iodide under certain circumstances. This could be explained by its higher hygroscopicity, the lower point of deliquescence and a smaller ionic size in comparison to potassium iodate. Different KI applications on butterhead lettuce during the course of the day showed the highest iodine concentration around midday and the wide iodine accumulation differences observed under fluctuating environmental conditions could be explained by the impact of light radiation, the relative humidity and the temperature conditions at the time of application. Furthermore, iodine treatments at different application dates nearing harvest showed a rising iodine concentration in butterhead lettuce that could be associated to the increasing fresh mass and to the rising leaf area.

The leaf topography of some vegetable crops was examined and the stomatal size and distribution density was found to influence the absorption of aerially applied iodine solutions. Moreover, iodine was estimated to be marginally phloem mobile.

The miscibility of iodine with some agrochemicals was positively tested and tank mixing adjudged to work well with other agrochemicals. When KI or KIO_3 were sprayed simultaneously with commercial calcium fertilizers, fungicides or insecticides, iodine accumulation in butterhead lettuce was not affected and in some cases was even significantly enhanced. The latter result was probably due to an optimized foliar spray formulation of the applied products.

Additionally, the application of iodine foliar sprays at a dose relevant for practical implementation was found to slightly enhance the content of the total phenolic compounds and thus may improve the nutritional quality of multi-leaf lettuce as well.

8 Zusammenfassung

Iod ist ein essentielles Element für den Menschen, das in vielen Regionen der Welt häufig nicht ausreichend mit der Nahrung aufgenommen wird. Trotz der verbreiteten Verwendung von iodiertem Speisesalz sind auch in Deutschland derzeit noch weite Teile der Bevölkerung mit diesem Spurenelement unterversorgt. Daher sind weitere Ansätze zur Verbesserung der alimentären Iod-Aufnahme erforderlich. Ziel dieser Arbeit war es, eine Verfahrenstechnik zu entwickeln, mit der Gemüse bereits beim Anbau im Feld bzw. Gewächshaus durch die Düngung von iodhaltigen Salzen gezielt mit diesem Mineralstoff angereichert werden kann. Hierzu wurden verschiedene methodische Ansätze (Boden- und Blattdüngung) bei ausgewählten Pflanzenarten miteinander verglichen sowie der Einfluss von Applikations-, Boden- und Umweltfaktoren auf die Effizienz dieser agronomischen Biofortifikation untersucht.

Die Iod-Biofortifikation verschiedener Gemüsearten mittels Boden- und Blattapplikationen wurde in einer Serie von experimentellen Freiland- und Gewächshausversuche über drei Anbaujahre (2010 - 2012) an mehreren Standorten in Niedersachsen und Nordrhein-Westfalen untersucht. Mit zunehmender Iod-Düngung des Bodens (0; 1,0; 2,5; 7,5 und 15 kg I ha^{-1}) in Form von KI und KIO_3 wurden steigende Iod-Konzentrationen in den essbaren Pflanzenteilen verzeichnet, wenn die Applikation direkt vor dem Pflanzen oder Säen erfolgte. Die höchsten Iod-Konzentrationen traten in Blatt- und Knollengemüsearten auf (Kopfsalat gefolgt von Eisbergsalat, Kohlrabi und Radieschen). In diesen Gemüsearten konnte der angestrebte Iod-Gehalt [≥ 50 µg I (100 g FM)$^{-1}$] bei KIO_3-Angebot mit einer Iod-Düngung in Höhe von 7,5 kg I ha^{-1} erreicht oder annähend erreicht werden, ohne dass dabei eine signifikante Ertragsreduktion oder Beeinträchtigungen in der Vermarktungsqualität zu beobachten waren. Bei einer Versorgung der Pflanzen mit KI in gleicher Höhe wurden hingegen eine generell viel geringere Iod-Anreicherung und bei manchen Gemüsearten zudem Ertragseinbußen in Höhe von über 10 % verzeichnet. Im Vergleich zu den zuvor genannten Gemüsearten reicherten Möhren und Zwiebeln nur wenig Iod in den essbaren Pflanzenteilen an.

Nach einer Bodendüngung mit KI und KIO_3 wurde das pflanzenverfügbare Iod im Wurzelraum sehr schnell in nicht-pflanzenverfügbaren Iod-Formen umgewandelt. Langzeiteffekte durch eine einmalige Iod-Düngegabe wurden dementsprechend nicht beobachtet. Der Vergleich zwischen Boden- und Blattapplikationen im ersten Versuchsjahr ergab bei Kopfsalat eine überlegene Effizienz der Blattspritzungen. Daher wurde dieser Biofortifikations-Ansatz in weiteren Feld- und Gewächshausversuchen an verschiedenen Gemüsearten näher untersucht. Mit zunehmender Iod-Düngung der oberirdischen Pflanzenteile (0; 0,1; 0,25; 0,5; 0,75; 1,0 und 2,0 kg I ha^{-1}) wurden, mit KI- oder KIO_3-Applikationen zu verschieden Zeitpunkten vor der Ernte, steigende Iod-Konzentrationen in den essbaren Pflanzenteilen festgestellt. Die höchsten Iod-Konzentrationen wurden in Blattgemüsearten (Winterspinat, Kopfsalat, Rucola, Basilikum, Petersilie und Oregano) ermittelt. Der gewünschte Iod-Gehalt konnte schon mit einer einmaligen Iod-Blattapplikation in geringen Dosen von 0,1 - 0,25 kg I ha^{-1} erreicht werden, ohne dass es dabei zu einer signifikante Ertragsreduktion oder einer Beeinträchtigung der Vermarktungsqualität kam. Blattapplikationen mit KI und KIO_3 führten unter kontrollierten Umweltbedingungen im Gewächshaus zu einer vergleichbar hohen Iod-Anreicherung wie im Freiland. Weitere Feldversuche mit Kopfsalat als Modellpflanze zeigten, dass bei Iodid unter gewissen Umweltbedingungen vom Blatt besser aufgenommen wird als Iodat. Dies ist vermutlich dadurch zu erklären, dass Iodid im Vergleich zu Iodat eine höhere Hygroskopizität, einen niedrigeren Deliqueszenzpunkt und einen kleineren Ionenradius besitzt. Mit KI als Düngesalz wurde die höchste Iod-Aufnahme bei einer Blattapplikation am späten Vormittag beobachtet. Diese Veränderungen im Tagesverlauf standen im Zusammenhang mit der Einstrahlung, der Lufttemperatur und der Luftfeuchte zum Applikationszeitpunkt. Des Weiteren wurde festgestellt, dass die Anreicherung des Spurenelements im geernteten Salatkopf umso höher war, je dichter die Iod-Spritzungen im Kulturverlauf am Erntetermin lagen. Dies konnte mit der Zunahme des Frischgewichts und einer damit einhergehenden erhöhten Blattfläche in Zusammenhang gebracht werden.

Auch die Blattbeschaffenheit einiger Gemüsearten wurde mikroskopisch untersucht und die Größe sowie die Dichte der Stomata als weitere mögliche Einflussfaktoren für eine Iod-Blattapplikation identifiziert. Darüber hinaus erwies sich Iod in den untersuchten Pflanzen als bedingt phloemmobil, so dass die Verlagerbarkeit des Mineralstoffs innerhalb der Pflanze (z.B. vom Blatt in Sprossknollen) begrenzt ist. Versuche zur Mischbarkeit von KI und KIO_3 mit Pflanzenschutz- und Düngemitteln haben eine gute Verträglichkeit der geprüften Komponenten gezeigt und die Iod-Anreicherung in den essbaren Pflanzenteilen von Kopfsalat war in manchen Fällen sogar erhöht. Dies ist vermutlich auf die optimierte Formulierung der zugegebenen Mittel zurückzuführen. Darüber hinaus wurde nach einer einmaligen Blattdüngung auf Multiblatt-Salaten eine leichte Erhöhung des Gesamtphenolgehalts festgestellt. Die Ergebnisse weisen darauf hin, dass die Iod-Biofortifikation über die Anreicherung des Iodgehaltes hinaus einen positiven Einfluss auf den gesundheitlichen Wert von Gemüse haben kann.

9 References

Abd El-Fattah, M. A. and E. M. R. Agwah 1987: Physiological studies on lettuce tipburn. Egyptian Journal of Horticulture, 14, 2, 143 – 153.

Alloway, B. J. 2008: Zinc in Soils and Crop Nutrition. Second edition, published by IZA and IFA, Brussels, Belgium and Paris, France.

Altinok, S., S. Sozudogru-Okand and H. Halilova 2003: Effect of Iodine Treatments on Forage Yields of Alfalfa. Communications in Soil Science and Plant Analysis, 34, 1, 55 – 64.

Amachi, S. 2008: Microbial Contribution to Global Iodine Cycling: Volatilization, Accumulation, Reduction, Oxidation, and Sorption of Iodine. Microbes Environ, 23, 4, 269 – 276.

Amachi, S., M. Kasahara, S. Hanada, Y. Kamagata, H. Shinoyama, T. Fujii and Y. Muramatsu 2003, Microbial participation in iodine volatilization from soils, Environ. Sci. Technol. 37, 3885 – 3890.

Andersson, M., V. Karumbunathan and M. B. Zimmermann 2012: Global iodine status in 2011 and trends over the past decade. Journal of Nutrition, 142, 4, 744 – 750.

Anke, M. and W. Arnhold 2008: Das Spurenelement Iod. Biol. Unserer Zeit, 38, 400 – 406.

Apelblat, A. and E. Korin 1998: The vapour pressures of saturated aqueous solutions of sodium chloride, sodium bromide, sodium nitrate, sodium nitrite, potassium iodate, and rubidium chloride at temperatures from 227 K to 323 K. J. Chem. Thermodynamics, 30, 59 – 71.

Arbeitskreis Jodmangel 2009: Daten und Fakten zum Stand des Jodmangels und der Jodversorgung in Deutschland. Jodversorgung aktuell, Ausgabe 2009. A. J. Organisationsstelle, Leimenrode 29, 60322 Frankfurt am Main.

Arbeitskreis Jodmangel 2013a: Jodmangel und Jodversorgung in Deutschland. Aktuelles zum derzeitigen Versorgungsstand und Handlungsbedarf. 4. Auflage, Stand: Januar 2013. A. J. Organisationsstelle, Leimenrode 29, 60322 Frankfurt am Main.

Arbeitskreis Jodmangel 2013b: Tabelle Versorgungslücke Jod. Accessed 12 January 2013, <http://jodmangel.de/servicematerial/>.

ARBUCKLE, K. 2012: Electrostatic sprayer to benefit smaller growers. Grapegrower and Winemaker, 587, 46.

ASARIA, P., D. CHISHOLM, C. MATHERS, M. EZZATI AND R. BEAGLEHOLE 2007: Chronic disease preven tion: health effects and financial costs of strategies to reduce salt intake and control tobacco use. Lancet, 370, 2044 – 2053.

ASHWORTH, D. J, L. LUO, R. XUAN AND S. R. YATES 2011: Irrigation, organic matter addition, and tarping as methods of reducing emissions of methyl iodide from agricultural soil. Environ. Sci. Technol. 45, 4, 1384 - 1390.

ASHWORTH, D. J. 2009: Transfers of iodine in the soil-plant-air-system: Solid-liquid partitioning, migration, plant uptake and volatilization, 107 – 118. In: V. Preedy, G. H. Burrow and R. R. Watson (Eds.): Comprehensive Handbook of Iodine – Nutritional, biochemical, pathological and therapeutic aspects. Academic Press, Amsterdam.

ASTM (AMERICAN SOCIETY FOR TESTING AND MATERIALS) 2012: Standard Practice for Evaluation of Physical Compatibility of Pesticides in Aqueous Tank Mixtures by the Dynamic Shaker Method. Active Standard ASTM E 1518 – 05. Book of Standards Volume: 11.05. ASTM International, West Conshohocken, PA, 2012. DOI: 10.1520/E1518-05R12.

AUMONT, G. AND J. C. TRESSOL 1986: Improved routine method for the determination of total iodine in urine and milk. Analyst, 111, 841 - 843.

BAN-NAI, T., Y. MURAMATSU AND S. AMACHI 2006: Rate of iodine volatilization and accumulation by filamentous fungi through laboratory cultures. Chemosphere, 65, 2216 – 2222.

BECK, M. 2009: Einfluss der Wasserspannung und des Bewässerungsverfahrens auf den Ertrag von Kopfsalat im Unterglasanbau. Informationsdienst Weihenstephan, Ausgabe April 2009. Staatliche Forschungsanstalt für Gartenbau Weihenstephan.

BECKER, B. R. AND B. A. FRICKE 1992: Transpiration and Respiration of Fruits and Vegetables. New Developments in Refrigeration for Food Safety and Quality, International Institute of Refrigeration, Paris, France, and American Society of Agricultural Engineers, St. Joseph, Michigan, 110 - 121.

BECKER, J. O., H. D. OHR, N. M. GRECH, M. E. MCGIFFEN AND J. J. SIMS 1998: Evaluation of Methyl Iodide as a Soil Fumigant in Container and Small Field Plot Studies. Pestic. Sci., 52, 58 - 62.

BENZIE, I. F. F. AND J. J. STRAIN 1996: The Ferric Reducing Ability of Plasma (FRAP) as a Measure of "Antioxidant Power": The FRAP Assay. Analytical Biochemistry, 239, 70 - 76.

BFR (BUNDESINSTITUT FÜR RISIKOBEWERTUNG) 2004: Nutzen und Risiken der Jodprophylaxe in Deutschland. Aktualisierte Stellungnahme des BfR vom 22. Juni 2004.

BFR (BUNDESINSTITUT FÜR RISIKOBEWERTUNG) 2007: Gesundheitliche Risiken durch zu hohen Jodgehalt in getrockneten Algen. Aktualisierte Stellungnahme Nr. 026/2007 des BfR vom 22. Juni 2004.

BFR (BUNDESINSTITUT FÜR RISIKOBEWERTUNG) 2012: Blutdrucksenkung durch weniger Salz in Lebensmitteln Stellungnahme Nr. 007/2012 des BfR, MRI und RKI vom 19. Oktober 2011.

BGVV (BUNDESINSTITUT FÜR GESUNDHEITLICHEN VERBRAUCHERSCHUTZ UND VETERINÄRMEDIZIN) 2001: Jodanreicherung von Lebensmitteln in Deutschland. Stellungnahme des Bundesinstitutes für gesundheitlichen Verbraucherschutz und Veterinärmedizin vom 05. Dezember 2001.

BIBBINS-DOMINGO, K., G. M. CHERTOW, P. G. COXSON, A. MORAN, J. M. LIGHTWOOD, M. J. PLETCHER AND L. GOLDMAN 2010: Projected Effect of Dietary Salt Reductions on Future Cardiovascular Disease. The New England Journal of Medicine, 362, 590 - 599.

BLANCO, A., V. FERNANDEZ AND J. VAL 2010: Improving the performance of calcium containing spray formulations to limit the incidence of bitter pit in apple (*Malus x domestica* borkh). Scientia Horticulturae, 127, 23 - 28.

BLASCO, B., J. J. RIOS, L. M. CERVILLA, E. SÁNCHEZ-RODRIGEZ, J. M. RUIZ AND L. ROMERO 2008: Iodine biofortification and antioxidant capacity of lettuce: potential benefits for cultivation and human health. Ann. App. Biol.152, 289 – 299.

BMEL (BUNDESMINISTERIUM FÜR ERNÄHRUNG UND LANDWIRTSCHAFT) 2014: Der Gartenbau in Deutschland, Daten und Fakten. Accessed 28 March 2014, <www.bmel.de/Publikationen>.

BMU (BUNDESMINISTERIUM FÜR UMWELT, NATURSCHUTZ UND REAKTORSICHERHEIT) 2004: Unterschiede bei der Ablagerung von Radionukliden auf verschiedene Blattgemüsearten. Schriften reihe Reaktorsicherheit und Strahlenschutz. BMU-2004-635.

BORST PAUWELS, G. W. F. H. 1961: Iodine as a micronutrient for plants. Plant and Soil, 14, 4, 377 – 392.

BOSTOCK, A. C., G. SHAW AND J. N. B. BELL 2003: The volatilization and sorption of ^{129}I in coniferous forest, grass-land and frozen soils. Journal of Environmental Radioactivity, 70, 29 - 42.

BÖSZÖRMÈNYI, Z., AND E. CSEH 1960: The uptake and reduction of iodate by wheat roots. Current Science 29, 340 - 341.

BOX, G. E. P. AND D. R. COX 1964: An Analysis of Transformations. Journal of the Royal Statistical Society, 26, 2, 211 - 252.

BRAVERMAN, L. E. 1994. Iodine and the thyroid - 33 years of study. Thyroid, 4, 351 - 356.

BROCKWELL, J., A. PILKA AND R. A. HOLLIDAY 1991: Soil pH is a major determinant of the numbers of naturally occurring *Rhizobium meliloti* in non-cultivated soils in central New South Wales. Australian Journal of Experimental Agriculture, 31, 2, 211 - 219.

BROWN, P. H. AND B. J. SHELP 1997: Boron mobility in plants. Plant and Soil, 193, 85 - 101.

BROWN, P. H. AND H. N. HU 1996: Phloem mobility of boron is species dependent: Evidence for phloem mobility in sorbitol-rich species. Annals of Botany, 77, 497 - 505.

BROWN, P. H. AND H. N. HU 1998: Phloem boron mobility in diverse plant species. Botanica Acta, 111, 331 - 335.

CABRAL, H. J. 2008: Multiple Comparisons Procedures. Circulation, 117, 698 - 701.

CAFFAGNI, A., N. PECCHIONI, P. MERIGGI, V. BUCCI, E. SABATINI, N. ACCIARRI, T. CIRIACI, L. PULCINI, N. FELICIONI, M. BERETTA AND J. MILC 2012: Iodine uptake and distribution in horticultural and fruit tree species. Italian Journal of Agronomy, 7, e32.

CAKMAK, I. 2008: Enrichment of cereal grains with zinc: agronomic or genetic biofortification? Plant Soil 302, 1 – 17.

CAKMAK, I., M. KALAYCI, Y. KAYA, A. A. TORUN, N. AYDIN, Y. WANG, Z. ARISOY, H. ERDEM, A. YAZICI, O. GOKMEN, L. OZTURK AND W. J. HORST 2010: Biofortification and localization of zinc in wheat grain. Journal of Agricultural and Food Chemistry, 58, 9092 - 9102.

CAMPBELL, C. AND C. O. PLANK 1998: Preparation of Plant Tissue for Laboratory Analysis, 37 - 49. In: Handbook of Reference Methods for Plant Analysis Edited by Y. P. Kalra. CRC Press, Lodon, New York, Washington, D. C.

CHAHAL, G. S., D. L. JORDAN, R. L. BRANDENBURG, B. B. SHEWC, J. D. BURTON, D. DANEHOWER AND A. C. YORK 2012: Interactions of agrochemicals applied to peanut; part 3: Effects on insect cides and prohexadione calcium. Crop Protection, 41, 150 - 157.

CHANG, X., P. G. ALDERSON AND C. J. WRIGHT 2008: Solar irradiance level alters the growth of basil (*Ocimum basilicum* L.) and its content of volatile oils. Environmental and Exper imental Botany, 63, 216 – 223.

CHAO, A., M. J. THUN, C. J. CONNELL, M. L. MCCULLOUGH, E. J. JACOBS, W. D. FLANDERS, C. RODRIGUEZ, R. SINHA AND E. E. CALLE 2005: Meat Consumption and Risk of Colorectal Cancer. Jour nal of American Medical Association, 293, 2, 172 - 182.

COSTA-FONT, M., J. M. GIL AND W. B. TRAILL 2008: Consumer acceptance, valuation of and attitudes towards genetically modified food: Review and implications for food policy. Food Policy, 33, 99 – 111.

D-A-CH 2000: Referenzwerte für die Nährstoffzufuhr. Deutsche Gesellschaft für Ernährung (DGE), Österreichische Gesellschaft für Ernährung (ÖGE), Schweizerische Gesellschaft für Ernährungsforschung (SGE), Schweizerische Vereinigung für Ernährung (SVE). Umschau Braus GmbH, Verlagsgesellschaft, Frankfurt a. M., 1. Auflage, 179 - 184.

D'AGOSTINO, R. B., A. BELANGER AND R. B. D'AGOSTINO JR. 1990: A Suggestion for Using Powerful and Informative Tests of Normality. The American Statistician, 44, 4, 316 - 321.

DAI, J.-L., M. ZHANG AND Y.-G. ZHU 2004b: Adsorption and desorption of iodine by various Chinese soils: I. Iodate. Environment International 4, 525 – 530.

DAI, J.-L., Y.-G. ZHU, M. ZHANG AND Y.-Z. HUANG 2004a: Selecting iodine-enriched vegetas and the residual effect of iodate application to soil. Biological Trace Element Research 3, 265 – 276.

DAI, J.-L., Y.-G. ZHU, Y.-Z. HUANG, M. ZHANG AND J. L. SONG 2006: Availability of iodide and iodate to spinach (*Spinacia oleracea* L.) in relation to total iodine in soil solution. Plant soil 289, 301 – 308.

DAVIS, D., J. CRAWFORD, S. LIU, S. MCKEEN, A. BANDY, D. THOMTON, F. ROWLAND AND D. BLAKE 1996: Potential impact of iodine on tropospheric levels of ozone and other critical oxidants. Journal of Geophysical Research, 101, D1, 2135 - 2147.

DELANGE, F. 1985: Physiopathology of iodine nutrition. In: Trace Elements in Nutrition of Children. R. K. Chandra (ed.), Nestlé Nutrition, Raven Press, Vevey, New York, 291 - 299.

DELANGE, F. 2002: Iodine deficiency in Europe and its consequences: an update. European Journal of Nuclear Medicine and Molecular Imaging, 29, 2 (Supplement), S404 - S416.

DELONG, G. R., P. W. LESLIE, S.-H. WANG, X.-M. JIANG, M.-L. ZHANG, M. A. RAKEMAN, J.-Y. JIANG, T. MA AND X.-Y. CAO 1997: Effect on infant mortality of iodination of irrigation water in a severely iodine-deficient area of China. The Lancet, 350, 9080, 771 - 773.

DIMMER, C., P. G. SIMMONDS, G. NICKLESS AND M. R. BASSFORD 2001: Biogenic fluxes of halomethanes from Irish peatland ecosystems. Atmospheric Environment, 35, 2, 321 – 330.

DIN (DEUTSCHES INSTITUT FÜR NORMUNG) 1986: DIN 38402-51, Mai 1986. Deutsche Einheitsver fahren zur Wasser-, Abwasser und Schlammuntersuchung; Kalibrierung von Analysenverfahren, Auswertung von Analysenergebnissen und lineare Kalibrierfunktionen für die Bestimmung von Verfahrenskenngrößen (A 51). Beuth Verlag, Berlin.

DIN (DEUTSCHES INSTITUT FÜR NORMUNG) 2000: DIN 19684-3, August 2000. Bodenuntersuchungsverfahren im Landwirtschaftlichen Wasserbau - Chemische Laboruntersuchungen - Teil 3: Bestimmung des Glühverlusts und des Glührückstands. Beuth Verlag, Berlin.

DIN (DEUTSCHES INSTITUT FÜR NORMUNG) 2004: DIN ISO 8466-2, Juni 2004. Wasserbeschaffenheit - Kalibrierung und Auswertung analytischer Verfahren und Beurteilung von Verfahrenskenndaten - Teil 2: Kalibrierstrategie für nichtlineare Kalibrierfunktionen zweiten Grades. Beuth Verlag, Berlin.

DIN (DEUTSCHES INSTITUT FÜR NORMUNG) 2005a: DIN ISO 10390, Dezember 2005. Bodenbeschaffenheit - Bestimmung des pH-Wertes. Beuth Verlag, Berlin.

DIN (DEUTSCHES INSTITUT FÜR NORMUNG) 2005b: DIN EN ISO 17294-2, Februar 2005. Wasserbeschaffenheit - Anwendung der induktiv gekoppelten Plasma-Massenspektrometrie (ICP-MS) - Teil 2: Bestimmung von 62 Elementen. Beuth Verlag, Berlin.

DIN (DEUTSCHES INSTITUT FÜR NORMUNG) 2007a: DIN EN 15111, Juni 2007. Lebensmittel - Bestimmung von Elementspuren - Bestimmung von Iod mit der ICP-MS (Massenspektrometrie mit induktiv gekoppeltem Plasma). Beuth Verlag, Berlin.

DIN (DEUTSCHES INSTITUT FÜR NORMUNG) 2007b: DIN EN ISO 16720, Juni 2007. Bodenbeschaffenheit - Vorbehandlung von Proben durch Gefriertrocknung für die anschließende Analyse. Beuth Verlag, Berlin.

DIOSADY, L. L., J. O. ALBERTI, K. RAMCHARAN AND M. G. VENKATESH MANNAR 2002: Iodine stability in salt double-fortified with iron and iodine. Food & Nutrition Bulletin, 23, 2, 196 - 207.

DUMONT, J. E., A. M. ERMANS, C. MAENHAUT, F. COPPÉE AND J. B. STANBURY 1995: Large goiter as a maladaption to iodine deficiency. Clinical Endocrinology, 43, 1 – 10.

DUNN, O. J. 1964: Multiple comparisons using rank sums. Technometrics, 6, 241 - 252.

DUNNETT, C. W. 1955: A Multiple Comparison Procedure for Comparing Several Treatments with a Control. Journal of the American Statistical Association, 50, 1096 - 1121.

DWD (DEUTSCHER WETTER DIENST) 2013: Accessed 23 March 2013, <http://www.dwd.de/>.

EFSA (EUROPEAN FOOD SAFETY AUTHORITY) 2006: Scientific Committee on Food - Scientific Panel on Dietetic Products, Nutrition and Allergies. European Food Safety Authority 2006. ISBN: 92-9199-014-0.

EICHERT, T. AND H. E. GOLDBACH 2008: Equivalent pore radii of hydrophilic foliar uptake routes in stomatous and astomatous leaf surfaces - further evidence for a stomatal pathway. Physiol. Plant, 132, 491 - 502.

EICHERT, T. AND J. BURKHARDT 2001: Quantification of stomatal uptake of ionic solutes using a new model system. Journal of Experimental Botany, 52, 771 - 781.

ENACADEMIC 2014: Goiter morphology - Struma nodosa. Accessed 14 March 2014, <http://en.academic.ru/dic.nsf/enwiki/7275>.

ENNIS, W. B. JR., R. E. WILLIAMSON AND K. P. DORSCHNER 1952: Studies on Spray Retention by Leaves of Different Plants. Weeds, 1, 3, 274 - 286.

EPSTEIN, E. AND A. J. BLOOM 2005: Mineral nutrition of plants: Principles and perspectives. 2nd Edition, Sinauer Associates, Inc.

ERDOGAN, M. F. 2003: Thiocyanate overload and thyroid disease. BioFactors 19, 107 – 111.

EU (EUROPEAN UNION) 2006: EU Regulation no 1924/2006 of The European Parliament and of the Council of 20 December 2006 on nutrition and health claims made on foods. Official Journal of the European Union, L 404/9.

EU (EUROPEAN UNION) 2011: EU Regulation no 1169/2011 of The European Parliament and of the Council of 25 October 2011 on the provision of food information to consumers. Official Journal of the European Union, L 304/18.

EU (EUROPEAN UNION) 2012: EU Regulation no 432/2012 of 16 May 2012 establishing a list of permitted health claims made on foods. Official Journal of the European Union, L 136/1.

FAO (THE FOOD AND AGRICULTURE ORGANIZATION OF THE UNITED NATIONS) 2013: The State of Food and Agriculture. FAO, Rome, 2013. ISBN 978-92-5-107671-2.

FAO/WHO (THE FOOD AND AGRICULTURE ORGANIZATION OF THE UNITED NATIONS/WORLD HEALTH ORGANIZATION) 2004: Human Vitamin and Mineral Requirements. Report of a joint FAO/WHO expert consultation, Bangkok, Thailand. Food and Nutrition Division, FAO, Rome.

FELLER, C., M. FINK, H. LABER, A. MAYNC, P. PASCHOLD, H. C. SCHARPF, J. SCHLAGHECKEN, K. STROHMEYER, U. WEIER AND J. ZIEGLER 2011: Düngung im Freilandgemüsebau. In: Fink, M. (Hrsg.): Schriftenreihe des Leibniz-Instituts für Gemüse- und Zierpflanzenbau (IGZ), 3. Auflage, Heft 4, Großbeeren.

FERNÁNDEZ, V. AND T. EICHERT 2009: Uptake of hydrophilic solutes through plant leaves: current state of knowledge and perspectives of foliar fertilization. Crit. Rev. Plant Sci., 28, 36 – 68.

FERNÁNDEZ, V., T. SOTIROPOULOS AND P. H. BROWN 2013: Foliar Fertilization: Principles and Practices. International Fertilizer Industry Association (IFA), Paris, France.

FOSS, O. P., L. HANKES AND D. D. VAN SILKE 1960: A study of the alkaline ashing method for determination of protein-bound iodine in serum. Clin Chim Acta, 5, 301 – 326.

FUGE, R. 1996. Geochemistry of iodine in relation to iodine deficiency diseases. Geological Society, London, Special Publications 113, 201 - 211.

FUGE, R. 2005. Soils and iodine deficiency. In: Essentials of Medical Geology. O. Selinus, B. Alloway, J. A. Centeno, R. B. Finkelman, R. Fuge, U. Lindh and P. Smedley (eds.), Elsevier, San Diego, CA, 417 – 433.

FUGE, R. 2007: Iodine deficiency: An ancient problem in a modern world. Ambio, 36, 70 - 72.

FUSE, H., H. INOUE, K. MURAKAMI, O. TAKIMURA AND Y. YAMAOKA 2003: Production of free and organic iodine by *Roseovarius* spp. FEMS Microbiology Letters, 229, 189 – 194.

GÄRTNER, R. 2000: Gibt es Risiken der Jodmangelprophylaxe? Ernährungs-Umschau, 47, 86 - 91.

GAUCH, H. G. JR. 1992: Statistical Analysis of Regional Yield Trials: AMMI Analysis of Factorial Designs. Elsevier Science & Technology.

GGP 2012: Iodí product line by Pizzoli S.p.A. Accessed 19 April 2013, <http://www.cgp.it/schede/scheda.php?idscheda=339#>.

GHASEMI, A. AND S. ZAHEDIASL 2012: Normality Tests for Statistical Analysis: A Guide for Non-Statisticians. Int J Endocrinol Metab., 10, 2, 486 - 489.

GOŁKOWSKI, F., Z. SZYBIŃSKI, J. RACHTAN, A. SOKOLOWSKI, M. BUZIAK-BEREZA, M. TROFIMIUK, A. HUBALEWSKA-DYDEJCZYK, E. PRZYBYLIK-MAZUREK AND B. HUSZNO 2007: Iodine prophylaxis – the protective factor against stomach cancer in iodine deficient areas. European Journal of Nutrition, 46, 5, 251 - 256.

GOMEZ, K. A. AND A. A. GOMEZ 1984: Statistical Procedures for Agricultural Research. Second Edition, John Wiley & Sons, Inc., U.K.

GRAHAM, P. H. 1981: Some problems of nodulation and symbiotic nitrogen fixation in *Phaseolus vulgaris* L.: a review. Field Crops Res., 4, 93 - 112.

GRAHAM, P. H., K. J. DRAGER AND M. L. FERREY 1994: Acid pH tolerance in strains of *Rhizobium* and *Bradyrhizobium*, and initial studies on the basis for acid tolerance of *Rhizobium tropici* UMR 1899. Can. J. Microbiol., 40, 198 - 207.

GRAHAM, P. H., S. E. VITERI, F. MACKIE, A. T. VARGAS AND A. PALACIOS 1982: Variation in acid soil tolerance among strains of *Rhizobium phaseoli*. Field Crops Res., 5, 121 - 128.

GRAY, T. 2010: Die Elemente - Bausteine unserer Welt. Komet Verlag GmbH, Köln.

GREER, M. A. 1957: Goitrogenic Substances in food. The American Journal of Clinical Nutrition, 5, 4, 440 - 444.

GREENSPAN, L. 1977: Humidity Fixed Points of Binary Saturated Aqueous Solutions. Journal of Research of the National Bureau of Standards-A. Physics and Chemistry. 81 A, 1, 89 - 96.

GRIFFITH, L. P. 2010: Grower Talks Magazine. Accessed 16 March 2013, <http://www.ballpublishing.com/growertalks/ViewArticle.aspx?articleid=17677>.

GRIMMINGER, S. P. 2005: Zum Iodbedarf und zur Iodversorgung der Haus- und Nutztiere und des Menschen. Dissertation Ludwig-Maximilian-Universität, München.

GROẞKLAUS, R. AND G. JAHREIS 2004: Universelle Salzjodierung für Mensch und Tier. Ernährungs-Umschau, 51, 138 - 143.

HAMPEL, R., J. KAIRIES AND H. BELOW 2009: Beverage iodine levels in Germany. Eur. Food Res. Technology 229, 705 – 708.

HARTEL, P. G. AND M. ALEXANDER 1983: Growth and survival of cowpea rhizobia in acid, aluminum-rich soils. Soil. Sci. Soc. Am. J., 47, 502 - 506.

HARTIKAINEN, H. 2005: Biogeochemistry of selenium and its impact on food chain quality and human health. J. Trace Elem. Med. Biol. 18: 309 – 311.

HEMMING, J., E. VAN OS AND J. BALENDONCK 2009: Intelligent bewässern im Gartenbau: Forschungstrends in den Niederlanden. Landbauforschung – Sonderh., 328, 81.

HERBISEED 2009: Featherlite side boom sprayer. Accessed 01 April 2014, <http://www.herbiseed.com/trials/home/sprayers/spraybooms/leaflet.aspx>.

HERRETT, R. A., H. H. HATFIELD JR., D. H. CROSBY AND A. J. VLITOS 1962: Leaf abscission induced by the iodide ion, Plant Physiol. 37, 358 – 363.

HETZEL, B. S., J. CHAVADEJ AND B. J. POTTER 1988: The Brain in Iodine deficiency. Neuropathology and Applied Neurobiology, 14, 2, 93 – 104.

HINTZE, J. 2007: NCSS Help System. Manual published by NCSS, Utah, U.S.A.

HINZE, G. AND J. KÖBBELING 1992: Alimentärer Jodmangel. Fortschr. Med. 110 163 – 166.

HIRSCHI, K. D. 2009: Nutrient Biofortification of Food Crops. Annual Review of Nutrition, 29, 1, 401 – 421.

HOCHBERG, Y. AND A. C. TAMHANE 1987: Multiple Comparison Procedures. John Wiley & Sons., New York.

HOLTSCHULZE, M. 2005: Tipburn in head lettuce - the role of calcium and strategies of preventing the disorder. Dissertation Rheinische Friedrich-Wilhelms-Universität zu Bonn.

HONG C.-L., H.-X. WENG, A.-L. YAN AND E.-U. ISLAM 2009: The fate of exogenous iodine in pot soil cultivated with vegetables. Environ. Geochem. Health, 1, 99 – 108.

HONG C.-L., H.-X. WENG, Y.-C. QIN, A.-L. YAN AND L.-L. XIE 2008: Transfer of iodine from soil to vegetables by applying exogenous iodine. Agron. Sustain. Dev., 28, 575 – 583.

HOU, X.-L., A. ALDAHAN, S. P. NIELSEN AND G. POSSNET 2009: Time Series of I-129 and I-127 Speciation in Precipitation from Denmark. Environmental Science & Technology, 43, 6522 - 6528.

HOU, X.-L., C. L. FOGH, J. KUČERA, K. G. ANDERSSON, H. DAHLGAARD AND S. P. NIELSEN 2003: Iodine-129 and Caesium-137 in Chernobyl contaminated soil and their chemical fractionation. Science of the Total Environment, 308, 97 - 109.

HSU, J. 1996: Multiple Comparisons: Theory and Methods. Chapman & Hall. London.

IGNAT, I., I. VOLF, V. I. POPA 2011: A critical review of methods for characterisation of polyphenolic compounds in fruits and vegetables. Food Chemistry, 126, 1821 – 1835.

IOM (INSTITUTE OF MEDICINE) 2006: Dietary Reference Intakes - The Essential Guide to Nutrient Intakes. The National Academic Press, Washington, D. C.

JENSEN, K. A., J. CONTI AND A. KJAER 1953: Isothiocyanates II: Volatile Isothiocyanates in Seeds and Roots of Various Brassicae. Acta Chemica Scandinavica, 7, 1267 - 1270.

JOHANSON, K. J. 2000: Iodine in soil. Department of Forest Mycology and Pathology. The Swedish University of Agricultural Sciences, Uppsala. Technical Report, TR-00-21.

JOHNER, A. S., A. L. B. GÜNTHER AND T. REMER 2011: Current trends of 24-h urinary iodine excretion in German schoolchildren and the importance of iodised salt in processed foods. Brith. J. Nutr. 106, 1749 – 1756.

JOHNER, A. S., K. VON NIDA, G. JAHREIS AND T. REMER 2012: Aktuelle Untersuchungen zeitlicher Trends und saisonaler Effekte des Jodgehaltes in Kuhmilch – Untersuchungen aus Nordrhein Westfalen. Berl Münch Tierärztl Wochenschr, 125, 76 – 82.

JOHNSON, C. C. 1980: The Geochemistry of Iodine and a Preliminary Investigation into its Potential Use as a Pathfinder Element in Geochemical Exploration. PhD thesis, University College of Wales, Aberystwyth.

JOHNSON, C. C. 2003: The geochemistry of iodine and its application to environmental strategies for reducing risks from iodine deficiency disorders. British Geological Survey Commissioned Report, CR/03/057N.

JOHNSON, C. C., M. H. STRUTT, M. HMEURRAS AND M. MOUNIR 2002: Iodine in the environment of the high atlas mountain area of Morocco. British geological survey, department for international development, DFID KAR Project R7411, Commissioned Report CR/02/196N.

JOPKE, P., J. FLECKENSTEIN, E. SCHNUG AND M. BAHADIR 1997: Spurenanalytik von Iod in Böden und Pflanzen. Analytiker-Taschenbuch, 15, 121 – 145.

JOPKE, P., M. BAHADIR, J. FLECKENSTEIN AND E. SCHNUG 1996: Iodine determination in plant materials. Commun. Soil Sci. Plant Anal., 27 , 741 - 751.

KATO, S., T. WACHI, K. YOSHIHIRA, T. NAKAGAWA, A. ISHIKAWA, D. TAKAGI, A. TEZUKA, H. YOSHIDA, S. YOSHIDA, H. SEKIMOTO AND M. TAKAHASHI 2013: Rice (*Oryza sativa* L.) roots have iodate reduction activity in response to iodine. Frontiers in Plant Science, 4, 227.

KEPPLER, F., R. EIDEN, V. NIEDAN, J. PRACHT AND H. F. SCHÖLER 2000: Halocarbons produced by natural oxidation processes during degradation of organic matter. Nature, 403, 298 - 301.

Kerschberger, M. and G. Franke 2001: Düngung in Thüringen nach „Guter fachlicher Praxis". Schriftenreihe Heft 11/2001, Thüringer Landesanstalt für Landwirtschaft, Jena.

Kerschberger, M., U. Hege and A. Jung 1997: Phosphordüngung nach Bodenuntersuchung und Pflanzenbedarf. VDLUFA Standpunkt, Verband Deutscher Landwirtschaftlicher Untersuchungs- und Forschungsanstalten, Darmstadt.

Knapp, G., B. Maichin, P. Fecher and S. Hasse 1998: Iodine determination in biological materials; Options for sample preparation and final determination. Fresenius Journal of Analytical Chemistry, 362, 508 - 513.

Knecht 2014: Irrigation gantries for the professionals. Accessed 10 January 2014, <http://www.knechtgmbh.com/download-archiv.html>.

Knorpp, L. and A. Kroke 2010: Salzreduktion als bevölkerungsbezogene Präventionsmaßnahme. Teil 1 – Handlungsbedarf in Deutschland. Ernährungs-Umschau, 57, 294 – 300.

Koch, H. 2007: So entstehen Spritzbelege. Pflanzenschutz-Praxis. DLG-Mitteilungen 02/2007, 66 - 69.

Kraemer, T., M. Hunsche and G. Noga 2009: Selected calcium salt formulations: Interactions between spray deposit characteristics and Ca penetration with consequences for rain-induced wash-off. J. Plant Nutr, 32, 1718 - 1730.

Krajčovičová-Kudláčková M., K. Bučková, I. Klimeš and E. Šeboková 2003: Iodine Deficiency in Vegetarians and Vegans. Ann. Nutr. Metab., 47, 183 – 185.

Kreeb, K. 1990: Methoden zur Pflanzenökologie und Bioindikation. Gustav Fischer Verlag, Stuttgart, New York.

Krohn, K. and R. Paschke 2009: Iodine deficiency, antioxidant response and mutagenesis in the thyroid gland: Antioxidant response and mutagenesis, 549 – 558. In: V. Preedy, G. H. Burrow and R. R. Watson (Eds.): Comprehensive Handbook of Iodine – Nutritional, biochemical, pathological and therapeutic aspects. Academic Press, Amsterdam.

KTBL (Kuratorium für Technik und Bauwesen in der Landwirtschaft) 2009: Gartenbau: Produktionsverfahren planen und kalkulieren. ISBN 978-3-9393-7179-3.

KUČERA, J. 2009: Assay of Iodine in foodstuffs: Methods and applications, 16 – 27. In: V. Preedy, G. H. Burrow and R. R. Watson (Eds.): Comprehensive Handbook of Iodine – Nutritional, biochemical, pathological and therapeutic aspects. Academic Press, Amsterdam.

KUČERA, J. AND I. KRAUSOVÁ 2007: Fast decomposition of biological and other materials for radiochemical activation analysis: A radiochemical study of element recoveries following alkaline-oxidative fusion. J. Radioanl. Nucl. Chem., 271, 577 - 580.

LANDINI, M., S. GONZALI AND P. PERATA 2011: Iodine biofortification in tomato. J. Plant Nutr. Soil Sci., 174, 480 - 486.

LANDINI, M., S. GONZALI, C. KIFERLE, M. TONACCHERA, P. AGRETTI, A. DIMIDA, P. VITTI, A. ALPI, A. PINCHERA AND P. PERATA 2012: Metabolic engineering of the iodine content in Arabidopsis. Sci. Rep., 2, 338.

LARABI-GRUET, N., A. CHAUSSE´, L. LEGRAND AND P. VITORGE 2007: Relative stabilities of Ce(IV) and Ce(III) limiting carbonate complexes at 5 – 50 ° C in Na^+ aqueous solutions, an electrochemical study. Electrochimica Acta, 52, 2401 – 2410.

LATSCHA, H. P. AND M. MUTZ 2011: Chemie der Elemente. Chemie-Basiswissen IV. 1. Aufl., Springer-Verlag, Berlin, Heidelberg.

LEHNARTZ 2011: Easy-Sprayer PLUS Bedienungsanleitung. C. A. Jürgen Lehnartz GmbH.

LEYVA, R., E. SÁNCHEZ-RODRÍGUEZ, J. J. RÍOS, M. M. RUBIO-WILHELMI, L. ROMERO, J. M. RUIZ AND B. BLASCO 2011: Beneficial effects of exogenous iodine in lettuce plants subjected to salinity stress. Plant Science, 181, 195 – 202.

LITTLE, T. M. AND F. J. HILLS 1978: Agricultural Experimentation - Design and Analysis. 1. Edition. John Wiley & Sons, ISBN 978-0-471-02352-4.

LIU, X.-F., S. ARDO, M. BUNNING, J. PARRY, K. ZHOU, C. STUSHNOFF, F. STONIKER, L. YU AND P. KENDALLA 2007: Total phenolic content and DPPH radical scavenging activity of lettuce (*Lactuca sativa* L.) grown in Colorado. LWT 40, 552 – 557.

MACKOWIAK, C. L. AND P. R. GROSSL 1999: Iodate and iodide effects on iodine uptake and partitioning in rice (*Oryza sativa* L.) grown in solution culture. Plant Soil 212, 135 – 143.

MANZ, F., M. ANKE, H. G. BOHNET, R. GÄRTNER, R. GROßKLAUS, M. KLETT AND R. SCHNEIDER 1998: Jod-Monitoring 1998: Repräsentative Studie zur Erfassung des Jodversorgungszustandes der Bevölkerung Deutschlands. Abschlußbericht des Forschungsvorhabens „Jod-Monitoring" des BMG. Schriftenreihe des Bundesministeriums für Gesundheit, Band 110, Baden-Baden, Nomos Verlagsgesellschaft.

MARSCHNER, P. 2012: Marschner's Mineral Nutrition of Higher Plants, 3rd ed., Academic Press/Elsevier, London.

MATTHES, W., R. FLUCHT AND M. STOEPPLER 1978: Automatisierung eines Verfahrens zur Jodbestimmung in Pflanzen- und Filterproben. Fresenius Z. Anal. Chem., 291, 217 - 220.

MCMILLAN, R. T. JR., H. H. BRYAN AND J. J. SIMS 2003: The effect of methyl iodide on *Rhizoctonia solani, Meloidogyne incognita*, Amaranth, Yellow Nutsedge, grasses and yield in tomato. Proc. Fla. State Hort. Soc., 116, 172 - 174.

MENG, W. AND P. C. SCRIBA 2002: Jodversorgung in Deutschland - Probleme und erforderliche Maßnahmen: Update 2002. Dtsch Ärztebl. 99, 2560 - 2564.

MERZ, W. AND W. PFAB 1969: Fünf Jahre organische Ultramikro-Elementaranalyse in einem Industrielaboratorium. In: Microchimica Acta, 57, 905 - 920.

MICHA, R., S. K. WALLACE AND D. MOZAFFARIAN 2010: Red and Processed Meat Consumption and Risk of Incident Coronary Heart Disease, Stroke, and *Diabetes mellitus*: A Systematic Review and Meta-Analysis. Circulation, 121, 2271 - 2283.

MIWA, H., I. AHMED, J. YOON, A. YOKOTA AND T. FUJIWARA 2008: *Variovorax boronicumulans* sp. *nov.*, a boron accumulating bacterium isolated from soil. International Journal of Systematic and Evolutionary Microbiology, 58, 286 – 289.

MORGAN, J. 2013: Biofortification - Lasting solutions to micronutrient malnutrition and world hunger. CSA News magazine. DOI: 10.2134/csa2013-58-1-1.

MUNNS, D. N. 1986: Acid soils tolerance in legumes and rhizobia. Advances in Plant Nutrition, 2, 63 - 91.

MUNNS, D. N. AND A. A. FRANCO 1982: Soil constraints to legume production, 133 - 152. In: BNF technology for tropical agriculture. Edited by P.H. Graham and S.C. Harris. Centro International de Agricultura Tropical, Cali, Colombia.

MURAMATSU, Y. AND S. AMACHI 2007: Awards Ceremony Speeches and Abstracts of the 17th Annual V. M. Goldschmidt Conference Cologne, Germany. Geochimica et Cosmo chimica Acta 71, 15, Supplement, A607 – A697.

MURAMATSU, Y. AND S. YOSHIDA 1995: Volatilization of methyl iodide from the soil-plant system. Atmospheric Environment, 29, 1, 21 - 25.

MURAMATSU, Y., D. CHRISTOFFERS AND Y. OHMOMO 1983: Influence of chemical forms on iodine uptake by plant. J. Radiat. Res. 24, 326 – 328.

MURAMATSU, Y., S. UCHIDA, M. SUMIYA, Y. OHMOMO AND H. OBATA 1989: Tracer experiments on transfer of radio-iodine in the soil-rice plant system. Water, Air, Soil Pollut., 45, 157 – 171.

MURAMATSU, Y., S. YOSHIDA, S. UCHIDA AND A. HASEBE 1996: Iodine desorption from rice paddy soil. Water Air and Soil Pollution, 86, 359 - 371.

MYNETT, A. AND R. L. WAIN 1971: Selective Herbicidal Activity of Iodide in Relation to Iodide Accumulation and Foliar Peroxidase Activity. Pestic. Sci. 2, 238 - 242.

MYNETT, A. AND R. L. WAIN 1973: Herbicidal Action of Iodide: Effect on Chlorophyll Content and Photosynthesis in Dwarf Bean *Phaseolus vulgaris*. Weed Res., 13, 101 - 109.

NMELVL (NIEDERSÄCHSISCHES MINISTERIUM FÜR ERNÄHRUNG, LANDWIRTSCHAFT, VERBRAUCHERSCHUTZ UND LANDESENTWICKLUNG) 2010: Die Landwirtschaft in Niedersachsen. Accessed 01 April 2012, <http://www.ml.niedersachsen.de>.

NEAL, C., M. NEAL, H. WICKHAM, L. HILL AND S. HARMAN 2007: Dissolved iodine in rainfall, cloud, stream and groundwater in the Plynhmon area of mid-Wales. Hydrology and Earth System Sciences, 11, 283 - 293.

O'KEEFE, J. H. AND L. CORDAIN 2004: Cardiovascular Disease Resulting From a Diet and Lifestyle at Odds With Our Paleolithic Genome: How to Become a 21st-Century Hunter-Gatherer. Mayo Clin Proc. 79, 101 - 108.

OESTLING, O., P. KOPP AND W. BURKKART 1989, Foliar uptake of caesium, iodine and strontium and their transfer to edible parts of beans, potatoes and radishes. Radiat. Phys. Chem., 33, 551 – 554.

OHR, H. D., J. J. SIMS, N. M. GRECH, J. O. BECKER AND M. E. MCGIFFEN JR. 1996: Methyl iodide, an ozone-safe alternative to methyl bromide as a soil fumigant. Plant Disease, 80, 731 - 735.

OSBORNE, W. 2010: Improving your data transformations: Applying the Box-Cox transformation. Practical Assessment, Research & Evaluation, 15, 12.

OTT, L. 1988: An introduction to statistical methods and data analysis, third edition, PWS-KENT Publishing Company, USA.

PAETZ, A. AND B. M. WILKE, 2005: Soil Sampling and Storage, 1, 1 - 46. In: Margesin, R. and Schinner, F. (Eds.): Manual of Soil Analyses – Monitoring and Assessing Soil Bioremediation.

PAHUJA, D. N., M. G. RAJAN, A. V. BORKAR AND A. M. SAMUEL 1993: Potassium iodate and its comparison to potassium iodide as a blocker of 131I uptake by the thyroid in rats. Health Phys., 65, 545 - 9.

PANDEY, K. B. AND S. I. RIZVI 2009: Plant polyphenols as dietary antioxidants in human health and disease. Oxidative Medicine and Cellular Longevity, 2, 5, 270 - 278.

PARKER, D. R. 2009: Perchlorate in the environment: the emerging emphasis on natural occurrence. Environ. Chem, 6, 1, 10 - 27.

PEARCE, S. C., G. M. CLARKE, G. V. DIKE AND R. E. KEMPSON 1988: A Manual of Crop Experimentation, Charles Griffin & Company Limited, London, U.K.

PFANNENSTIEL, P. AND B. SALLER 1991: Schilddrüsenkrankheiten: Diagnose und Therapie, 2. Auflage, Berliner Medizinische Verlagsanstalt, Berlin.

POLYAK, D. E. 2013: Iodine. In: U.S. Geological Survey, 2013, Mineral commodity summaries 2013. USGS, ISBN 978–1–4113–3548–6.

POORTER, H. AND C. REMKES 1990: Leaf area ratio and net assimilation rate of 24 wild species differing in relative growth rate. Oecologia, 83, 553 - 559.

RANGASWAMY, R. 2010: A Textbook of Agricultural Statistics. Second edition, New Age International (P) Ltd., New Delhi.

RAVENSDOWN 2013: Stock Iodine 5%, Accessed 20 January 2014, <http://www.ravensdown.co.nz/nz/products/pages/animal-health/minerals-and-vitamins/stock-iodine.aspx>.

REDEKER, K. R., N. Y. WANG, J. C. LOW, A. MCMILLAN, S. C. TYLER AND R. J. CICERONE 2000: Emission of methyl halides and methane from rice paddies. Science, 290, 966 - 968.

REMER, T., A. NEUBERT AND F. MANZ 1999: Increased risk of iodine deficiency with vegetarian nutrition. British Journal of Nutrition, 81, 45 – 49.

REMER, T., S. JOHNER AND M. THAMM 2012: Jodversorgung von Schulkindern in Deutschland - Ergebnisse der DONALD-Studie. Deutsche Gesellschaft für Ernährung (Ed.), 12. Ernährungsbericht.

REN Q., F. FAN, Z. ZHANG, X. ZHENG, G.R. AND DELONG 2008: An environmental approach to correcting iodine deficiency: Supplementing iodine in soil by iodination of irrigation water in remote areas. Journal of Trace Elements in Medicine and Biology, 22, 1 – 8.

RICHARDSON, D. P. 2007: Risk management of vitamins and minerals. Food Science and Technology Bulletin: Functional Foods, 4, 6, 51 – 66.

RIEDERER, M., AND A. FRIEDMANN 2006: Transport of lipophilic non-electrolytes across the cuticle. In: Riederer, M. and C. Müller (eds.), Biology of the Plant Cuticle. Annual Plant Reviews, 23, 250 – 279. Blackwell Publishing, Oxford.

RIJK-ZWAAN 2009: RZ Seeds & Services Chainmail August 2009. Accessed 11 April 2013, <http://www.rijkzwaan.de/wps/wcm/connect/22d05a8042f46f67a16ceb80764 9d3c0/S%26S_Chainmail_August09.pdf?MOD=AJPERES>.

RIJK-ZWAAN 2012: RZ Seeds & Services März 2012. Accessed 11 April 2013, <http://www.rijkzwaan.de/wps/wcm/connect/56c732804a7f933fa5c5b57f681 92162/RZD-0991_Seeds_Services_03_2012_web-gesammt.pdf?MOD=AJPERES>.

ROTI, E. AND E. DEGLI UBERTI 2001: Iodine Excess and Hyperthyroidism. Thyroid, 11, 5.

RUEEGG, J., R. EDER AND V. ANDERAU 2006: Improved application Techniques - Ways to Higher Efficacy of Fungicides and Insecticides in Field Grown Vegetables. Outlooks on Pest Management, 17, 2, 80 - 84.

SANDELL, E. B. AND I. M. KOLTHOFF 1934: Chronometric catalytic method for the determination of micro-quantities of iodine. J. Am. Chem. Soc., 56, 1426 - 1435.

SANDELL, E. B. AND I. M. KOLTHOFF 1937: Micro determination of iodine by a catalytic method. Microchim. Acta, 1, 9 - 25.

SASTRY, S. K., C. D. BAIRD AND D. E. BUFFINGTON 1978: Transpiration Rates of Certain Fruits and Vegetables. ASHRAE Transactions, 84, 1, 237 - 254.

SATOLA, B., J. H. WÜBBELER AND A. STEINBÜCHEL 2013: Metabolic characteristics of the species *Variovorax paradoxus*. Appl. Microbiol. Biotechnol., 97, 541 – 560.

SCF (SCIENTIFIC COMMITTEE ON FOOD) 2002: Opinion of the Scientific Committee on Food on the tolerable upper intake level of Iodine. SCF/CS/NUTUPPLEV/26 final.

SCHLEGEL, T. K., J. SCHÖNHERR AND L. SCHREIBER 2006: Rates of foliar penetration of chelated Fe (III): Role of light, stomata, species, and leaf age. Journal of Agricultural and Food Chemistry, 54, 6809 - 6813.

SCHMITZ-EIBERGER, M. A., R. HAEFS AND G. J. NOGA 2002: Enhancing biological efficacy and rainfastness of foliar applied calcium chloride solutions by addition of rapeseed oil surfactants. Journal of Plant Nutrition and Soil Science, 165, 634 - 639.

SCHNELL, D. AND D. C. AUMANN 1999: The origin of iodine in soil: II. iodine in soil of Germany, Chemie der Erde: Beiträge zur chemischen Mineralogie, Petrographie und Geologie, 59, 69 -76.

SCHÖNHERR, J. 2000: Calcium chloride penetrates plant cuticles via aqueous pores. Planta, 212, 112 - 118.

SCHÖNHERR, J. 2001: Cuticular penetration of calcium salts: Effects of humidity, anions, and adjuvants. Journal of Plant Nutrition and Soil Science, 164, 225 - 231.

SCHÖNINGER, W. 1955: Eine mikroanalytische Schnellbestimmung von Halogen in organischen Substanzen. Microchim. Acta, 43, 123 - 129.

SCHREIBER, L. AND J. SCHÖNHERR 2009: Water and Solute Permeability of Plant Cuticles - Measurement and Data Analysis, Springer-Verlag, Berlin, Heidelberg.

SCHÜCKING, B. AND S. RÖHL 2010: Effektivität der Beratung zur Jodsubstitution bei Schwangeren. Abschlussbericht zum BMBF-Forschungsprojekt 01EL0410.

SCHUMM-DRÄGER P. M. AND J. FELDKAMP 2007: Schilddrüsenkrankheiten in Deutschland Ausmaß, Entwicklung, Auswirkungen auf das Gesundheitswesen und Präventionsfolge. Prävention und Gesundheitsförderung 2, 153 – 158.

SCRIBA, P. C., H. HESEKER AND A. FISCHER 2007: Jodmangel und Jodversorgung in Deutschland: Erfolgreiche Verbraucherbildung und Prävention am Beispiel von jodiertem Speisesalz. Präv. Gesundheitsf., 2, 143 – 148.

SHAHRIARI, Z., B. HEIDARI, M. CHERAGHI AND A. G. SHAHRIARI 2013: Biofortification of staple food crops: Engineering the metabolic pathways. International Research Journal of Applied and Basic Sciences, 5, 3, 287 - 290.

SHEPPARD, M. I. AND D. H. THIBAULT 1991: A four-year mobility study of selected trace elements and heavy metals. J. Environ. Qual., 20, 101 - 114.

SHEPPARD, M. I. AND D. H. THIBAULT 1992: Chemical Behavior of Iodine in Organic and Mineral Soils. Applied Geochemistry, 7, 265 - 272.

SHEPPARD, M. I., D. H. THIBAULT, J. MCMURRY AND P. A. SMITH 1995: Factors affecting the soil sorption of iodine. Water, Air and Soil Pollution 83, 51 – 67.

SHEPPARD, M. I., M. MOTYCKA AND P. A. SMITH 1997: Soil Sorption of Iodine: Effects of pH and Enzimes. AECL EACL, Technical Record TR-777, COG-97-25-I.

SHEPPARD, S. C. AND W. G. EVENDEN 1992: Response of some vegetable crops to soil-applied halides. Canad. J. Soil Sci. 72, 555 - 567.

SHETAYA, W. H. A. H. 2011: Iodine Dynamics in Soil. PhD thesis, University of Nottingham.

SHIMAMOTO, Y. S., Y. TAKAHASHI AND Y. TERADA 2011: Formation of Organic Iodine Supplied as Iodide in a Soil–Water System in Chiba, Japan. Environ. Sci. Technol., 45, 6, 2086 – 2092.

SHINONAGA, T., M. H. GERZABEK, F. STREBL AND Y. MURAMATSU 2001: Transfer of iodine from soil to cereal grains in agricultural areas of Austria, Science of the Total Environment, 267, 33 – 40.

SINGELTON, V. L., R. ORTHOFER AND R. M. LAMUELA-RAVENTOS 1999: Analysis of total phenols and other oxidation substrates and antioxidants by means of Folin-Ciocalteu reagent. Methods Enzymol., 299, 152 - 178.

SINGH, P. N., J. SABATÉ AND G. E. FRASER 2003: Does low meat consumption increase life expectancy in humans? Am J Clin Nutr, 78 (suppl.), 526S – 532S.

SINGHAL, R. K., U. NARAYANAN AND I. BHAT 1998: Investigations on interception and translocation for airborne ^{85}Sr, ^{131}I and ^{137}Cs in beans, spinach and radish plants. Water, Air and Soil Pollution 101, 163 – 176.

SINGLETON, V. L. AND J. A. ROSSI JR. 1965: Colorimetry of Total Phenolics with Phosphomolybdic-Phosphotungstic Acid Reagents. Am. J. Enol. Vitic, 16, 3, 144 - 158.

SLIESARAVIČIUS, A., J. PEKARSKAS, V. RUTOVIENĖ AND K. BARANAUSKIS 2006: Grain yield and disease resistance of winter cereal varieties and application of biological agent in organic agriculture. Agronomy research, 4 (Special issue), 371 - 378.

SMOLEŃ, S., R. RAKOCZY, Ł. SKOCZYLAS, J. WIERZBIŃSKA, W. SADY AND I. LEDWOŻYW-SMOLEŃ 2013: Double Biofortification with Iodine and Selenium of Lettuce Cultivated in Field – Preliminary Analysis of the Problem. In: XVII. IPNC Proceeding Book, 956 - 957. Sabanci University, Istanbul.

SMOLEŃ, S., S. ROZEK, I. LEDWOZYW-SMOLEN AND P. STRZETELSKI 2011a: Preliminary evaluation of the influence of soil fertilization and foliar nutrition with iodine on the efficiency of iodine biofortification and chemical composition of lettuce. J. Element. 16, 4, 613 – 622.

SMOLEŃ, S., S. ROZEK, P. STRZETELSKI AND I. LEDWOZYW-SMOLEŃ 2011b: Preliminary evaluation of the influence of soil fertilization and foliar nutrition with iodine on the effectiveness of iodine biofortification and mineral composition of carrot. J. Elementol., 16, 1, 103 – 114.

SOMERS, T. C., AND G. ZIEMELIS 1985: Spectral evaluation of total phenolic components in Vitis vinifera: Grapes and wines. Journal of the Science of Food and Agriculture 36, 12, 1275 – 1284.

SONG, Y., J. E. MANSON, J. E. BURING AND L. SIMIN 2004: A Prospective Study of Red Meat Consumption and Type 2 Diabetes in Middle-Aged and Elderly Women. Diabetes Care, 27, 2108 – 2115.

SOUCI, S. W., W. FACHMANN AND H. KRAUT 2000: Die Zusammensetzung der Lebensmittel. Nährwert-Tabellen. 6. Aufl. Medpharm Scientific Publishers, Stuttgart.

SQM 2013: Applications of Iodine. Accessed 22 December 2013, <http://www.sqm.com/en-us/home.aspx>.

STATISTISCHES BUNDESAMT 2011: Statistisches Jahrbuch für die Bundesrepublik Deutschland 2011. Statistisches Bundesamt, Wiesbaden, Kapitel 9, Gesundheitswesen.

STATISTISCHES BUNDESAMT 2013: Erhebungen zum Gemüseanbau in Deutschland neu konzipiert. Auszug aus Wirtschaft und Statistik, Statistisches Bundesamt, Wiesbaden 2013.

STEIDLE, B. 1989 Iodine-induced hyperthyroidism after contrast media. Animal experimental and clinical studies. In: Taenzer V, Wend S (eds) Recent Developments in Non-Ionic Contrast Media. Thieme, New York.

STEINBERG, S. M. G. M. KIMBLE, G. T. SCHMETT, D. W. EMERSON,M. F. TURNER AND M. RUDIN 2008: Abiotic reaction of iodate with sphagnum peat and other natural organic matter. Journal of Radioanalytical and Nuclear Chemistry, 277, 185 - 191.

STOCK, D. AND P. J. HALLOWAY 1993: Possible Mechanisms for Surfactant-Induced Foliar Uptake of Agrochemicals*. Pestic. Sci., 38, 165 - 177.

STOEWSAND, G. S. 1995: Bioactive organosulfur phytochemicals in *Brassica oleracea* vegetables - a review. Food Chem Toxicol, 33, 6, 537 - 43.

STRANG, J., T. JONES, R. BESSIN, B. ROWELL AND W. DUNWELL 1997: Integrated Crop Management for Kentucky Cabbage. Scout Manual. IPM-11.

STRZETELSKY, P., S. SMOLEŃ, S. ROZEK AND W. SADY 2010: Effect of Differentiated Fertilization and Foliar Application of Iodine on Yielding and Antioxidant Properties in Radish (*Raphanus sativus L.*) Plants. Ecological Chemistry and Engineering A, 17, 9, 1189 - 1195.

SUZUKI, M., E. YOSHIFUMI, S. OHSAWA, Y. KANESAKI, H. YOSHIKAWA, K. TANAKA, Y. MURAMATSU, J. YOSHIKAWA, I. SATO, T. FUJII AND SEIGO AMACHI 2012: Iodide Oxidation by a Novel Multicopper Oxidase from the Alphaproteobacterium Strain Q-1. Applied and Environmental Microbiology, 78, 11, 3941 – 3949.

SWAIN, P. A. 2005: Bernard Courtois (1777-1838), famed for discovering iodine (1811), and his life in Paris from 1798. Bull. Hist. Chem., 30, 2.

SWITALA, K. 2001: Determination of Iodide in 0.2 M Potassium Hydroxide by Flow Injection Analysis. QuikChem® Method 10-136-09-1-A, Lachat Instruments.

SZWONEK, E. 2009: Impact of Foliar Fertlizer Containing Iodine on "Golden Delicius" Apple Trees. In: The Proceedings of the International Plant Nutrition Colloquium XVI. UC Davis. Accessed 07 April 2012, <http://escholarship.org/uc/item/8bp5w7z7>.

TAIZ, L. AND E. ZEIGER 2006: Plant Physiology. 4th Edition, Sinauer Associates, Inc.

TANAKA. K., Y. TAKAHASHI, A. SAKAGUCHI, M. UMEO, S. HAYAKAWA, H. TANIDA, T. SAITO AND Y. KANAI 2012: Vertical profiles of Iodine-131 and Cesium-137 in soils in Fukushima Prefecture related to the Fukushima Daiichi Nuclear Power Station Accident. Geochemical Journal, 46, 73 - 76.

TAPPEL, A. 2007: Heme of consumed red meat can act as a catalyst of oxidative damage and could initiate colon, breast and prostate cancers, heart disease and other diseases. Med. Hypotheses, 68, 3, 562 - 564.

THAMM, M., U. ELLERT, W. THIERFELDER, K.-P. LIESENKÖTTER AND H. VÖLZKE 2007: Jodversorgung in Deutschland – Ergebnisse des Jodmonitorings im Kinder- und Jugendgesundheitssurvey (KiGGS). Bundesgesundheitsblatt Gesundheitsforschung Gesundheitsschutz, 50, 744 – 749.

TIWARI, R. P., W. G. REEVE, M. J. DILWORTH AND A. R. GLENN: Acid tolerance in *Rhizobium meliloti* strain WSM419 involves a two-component sensor-regulator system. Microbiobgy, 142, 1693 - 1704.

TONACCHERA, M., A. DIMIDA, M. DE SERVI, M. FRIGERI, E. FERRARINI, G. DE MARCO, L. GRASSO, P. AGRETTI, P. PIAGGI, F. AGHINI-LOMBARDI, P. PERATA, A. PINCHERA AND P. VITTI 2013: Iodine Fortification of Vegetables Improves Human Iodine Nutrition: In Vivo Evidence for a New Model of Iodine Prophylaxis. J Clin Endocrinol Metab, 98, E694 - E697.

TONACCHERA, M., A. PINCHERA, A. DIMIDA, E. FERRARINI, P. AGRETTI, P. VITTI, F. SANTINI, K. CRUMP AND J. GIBBS 2004: Relative Potencies and Additivity of Perchlorate, Thiocyanate, Nitrate, and Iodide on the Inhibition of Radioactive Iodide Uptake by the Human Sodium Iodide Symporter. Thyroid, 14, 12, 1012 - 1019.

TRUESDALE W. AND D. JONES 1996: The variation of iodate and total iodine in some UK rain waters during 1980-1981. J. Hydrol., 179, 1 - 4, 67 – 86.

TSUKADA, H., A. TAKEDA, K. TAGAMI AND S. UCHIDA 2008: Uptake and distribution of iodine in rice plants. J. Environ. Qual. 37, 2243 – 2247.

USDA (UNITED STATES DEPARTMENT OF AGRICULTURE) 2011: National Nutrient Database for Standard Reference. 11250, Lettuce, butterhead (includes boston and bibb types), raw. Accessed 17 April 2014, <http://ndb.nal.usda.gov/>.

VAN DE ZANDE, J. C., C. S. PARKIN AND A. J. GILBERT 2003: Application Technologies, 23 - 43. In: Wilson, M. F. (Ed.): Optimising Pesticide Use, John Wiley and Sons Ltd, England.

VARGAS, A. A. T. AND GRAHAM, P. H. 1988: *Phaseolus vulgaris* cultivar and *Rhizobium* strain variation in acid-pH tolerance and nodulation under acid conditions. Field Crops Res., 19, 91 - 101.

VDLUFA (VERBAND DEUTSCHER LANDWIRTSCHAFTLICHER UNTERSUCHUNGS- UND FORSCHUNGSANSTALTEN) 1997: Methodenbuch Band I. Die Untersuchung von Böden. Vierte Auflage. 1. Teillieferung 1991. 2. Teillieferung 1997. VDLUFA-Verlag, Darmstadt.

VERKAIK-KLOOSTERMAN, J., M. T. MCCANN, J. HOEKSTRA AND H. VERHAGEN 2012: Vitamins and minerals: issues associated with too low and too high population intakes. Food & Nutrition Research, 56, 5728.

VERMA, P, K. V. GEORGE, H. V. SINGH AND R. N. SINGH 2007: Modeling cadmium accumulation in radish, carrot, spinach and cabbage. Applied Mathematical Modelling, 31, 1652 – 1661.

VOOGT, W. 2005: Fertigation in Greenhouse Production. International Symposium on Fertigation. Optimizing the utilization of water and nutrients. Beijing, September 20 - 24, 2005. In: Fertigation: Optimizing the Utilization of Water and Nutrients. International Potash Institute, Horgen, Switzerland, 2008.

VOOGT, W. AND W. A. JACKSON 2010: Perchlorate, Nitrate, and Iodine Uptake and Distribution in Lettuce J. Agric. Food Chem., 58, 23, 12192 – 12198.

VOOGT, W., H. T. HOLWERDA AND R. KHODABAKS 2010: Biofortification of lettuce (*Lactuca sativa* L.) with iodine: the effect of iodine form and concentration in the nutrient solution on growth, development and iodine uptake of lettuce grown in water culture. Journal of the Science of Food and Agriculture, 90, 906 - 913.

WATERHOUSE, A. L. 2002: Wine phenolics. Ann N Y Acad Sci., 957, 21 - 36.

WEHRMANN, J. AND H. C. SCHARPF 1986: The N_{min}-method - an aid to integrating various objectives of nitrogen fertilization. Z. Pflanzenern. Bodenk. 149, 428 - 440.

WELCH, R. M. AND GRAHAM, R. D. 2004: Breeding for micronutrients in staple food crops from a human nutrition perspective. Journal of Experimental Botany, 55, 396, 353 - 364.

WELSHMAN, S. G., J. F. BELL AND G. MCKEE 1966: Automated method for the estimation of protein-bound iodine following alkaline incineration. J. Clin. Path., 19, 510.

WENG, H.-X, J. K. WENG, A.-L. YAN, C.-L. HONG, W.-B. YONG AND Y.-C. QIN 2008a: Increment of iodine content in vegetable plants by applying iodized fertilizer and the residual characteristics of iodine in soil, Biol. Trace Elem. Res., 123, 218 – 228.

WENG, H.-X., A.-L. YAN, C.-L. HONG, L.-L. XIE, Y. C. QIN AND C. Q. CHENG 2008b: Uptake of different species of iodine by water spinach and its effect to growth. Biol. Trace Elem. Res., 124, 184 – 194.

WENG, H.-X., A.-L. YAN, C.-L. HONG, Y.-C. QIN, L. PAN AND L.-L. XIE 2009: Biogeochemical transfer and dynamics of iodine in a soil-plant system, Environ. Geochem. Health 31, 401 – 411.

WENG, H.-X., C.-L. HONG AND A. L. YAN 2013: Biogeochemical transport of iodine and its quantitative model. Science China Earth Sciences, 56, 9, 1599 - 1606.

WENG, H.-X., C.-L. HONG, T.-H. XIA, L.-T. BAO, H.-P. LIU AND D.-W. LI 2012: Iodine biofortification of vegetable plants — An innovative method for iodine supplementation. Chin Sci Bull, 58, 17, 2066 - 2072.

WENG, H.-X., H.-P. LIU, D.-W. LI, M. YE, L. PAN AND T.-H. XIA 2014: An innovative approach for iodine supplementation using iodine-rich phytogenic food. Environ. Geochem. Health. DOI 10.1007/s10653-014-9597-4.

WHITE, P. J. AND M. R. BROADLEY 2009, Biofortification of crops with seven mineral elements often lacking in human diets – iron, zinc, copper, calcium, magnesium, selenium and iodine, New Phytol., 182, 49 – 84.

WHITE, P. J. AND M. R. BROADLEY, 2005: Biofortifying crops with essential mineral elements. Trends Plant Sci., 10, 586 - 593.

WHITEHEAD, D. C. 1973: Uptake and Distribution of Iodine in Grass and Clover Plants Grown in Solution Culture. J. Sci. Fd Agric., 24, 43 – 50.

Whitehead, D. C. 1975: Uptake by perennial ryegrass of iodide, elemental iodine and iodate added to soil as influenced by various amendments. Journal of the Science of Food and Agriculture, 26, 361 - 367.

Whitehead, D. C. 1978: Iodine in Soil Profiles in Relation to Iron and Aluminum-Oxides and Organic-Matter. Journal of Soil Science, 29, 88 - 94.

Whitehead, D. C. 1981: The volatilization, from soils and mixtures of soil components, of iodine added as potassium iodide. Journal of Soil Science, 32, 97 – 102.

Whitehead, D. C. 1984: The distribution and transformations of iodine in the environment. Environ. Int., 10, 321 – 339.

Whitten, K. W., Davis R. E., Peck M. L. and Stanley G. G. 2007: Chemistry, Eight Edition, Thomson Brooks/Cole, U.S.A.

WHO (World Health Organization) 2004: Iodine status worldwide. Department of Nutrition for Health and Development, World Health Organization. 20 Avenue Appia, 1211 Geneva 27, Switzerland.

Willacy 2012: Woman with a large multinodular goiter. Accessed 20 January 2014, <http://www.patient.co.uk/doctor/Thyroid-Lumps.htm>.

Wonneberger, C., F. Keller, H. Bahnmüller, H. Böttcher, B. Geyer and J. Meyer 2004: Gemüsebau. Eugen Ulmer GmbH & Co., Stuttgart.

Yamaguchi, N., M. Nakano, H. Tanida, H. Fujiwara and N. Kihou 2006: Redox reaction of iodine in paddy soil investigated by field observation and the IK-edge XANES fingerprinting method. Journal of Environmental Radioactivity, 86, 212 - 226.

Yara 2014a: Livestock - Iodine (I) - Grasstrac. Accessed 05 March 2014, <http://www.megalab.net/content/prodlivestockhomeigrasstrac.aspx>.

Yara 2014b: Livestock - Iodine (I) – Foliar Grasstrac. Accessed 05 March 2014, < http://www.megalab.net/content/prodlivestockhomeigrasstracfoliar.aspx>.

Yuita, K. 1992: Dynamics of Iodine, Bromine, and Chlorine in Soil. II: Chemical Forms of Iodine in Soil Solutions. Soil Science and Plant Nutrition, 38, 281 - 287.

YUITA, K., N. KIHOU, S. YABUSAKI, Y. TAKAHASHI, T. SAITOH, A. TSUMURA AND H. ICHIHASHI 2005: Behavior of Iodine in a Forest Plot, an Upland Field and a Paddy Field in the Upland Area of Tsukuba, Japan. Iodine Concentration in Precipitation, Irrigation Water, Ponding Water and Soil Water to a Depth of 2.5 m. Soil Sci Plant Nutr, 51, 1011 – 1021.

YUN, A. J. AND J. D. DOUX 2009: Iodine in the Ecosystem: An Overview, 119 – 123. In: V. Preedy, G. H. Burrow and R. R. Watson (Eds.): Comprehensive Handbook of Iodine – General Aspects of Iodine Sources and Intakes. Academic Press, Amsterdam.

ZANIRATO, V. 2008: International Patent PCT/EP2008/052455: "Method and composition for enriching potatoes with iodine and potatoes obtained thereby". Applicant: Pizzoli S.p.A.

ZHANG, Y., G. ZHANG AND F. HAN 2006: The spreading and superspreading behavior of new glucosamide-based trisiloxane surfactants on hydrophobic foliage. Colloids and Surfaces A: Physicochem. Eng. Aspects, 276, 100 – 106.

ZHAO, F.-J. AND P. R. SHEWRY, 2011: Recent developments in modifying crops and agronomic practice to improve human health. Food Policy, 36, 94 - 101.

ZHU, C., S. NAQVI, S. GOMEZ-GALERA, A. M. PELACHO, T. CAPELL, P. CHRISTOU 2007: Transgenic strategies for the nutritional enhancement of plants. Trends Plant Sci. 12, 548 – 55.

ZHU, Y.-G., Y.-Z. HUANG, Y. HU AND Y.-X. LIU 2003: Iodine uptake by spinach (*Spinacia oleracea* L.) plants grown in solution culture: effects of iodine species and solution concentrations. Environment International, 29, 33 – 37.

Acknowledgements

I want to express my sincere thanks to the following persons, companies and institutions:

Prof. Dr. D. Daum, initiator and leader of the iodine biofortification project at the University of Applied Sciences of Osnabrück; who always supported and critically advised me with dedication and a smile.

Prof. Dr. habil. H. Meuser and Prof. Dr. habil. J. W. Härtling from the Institute of Geography at the University of Osnabrück; for their valuable suggestions and the evaluation of this dissertation.

My dear colleague Dipl. Ing. (FH) R. Czauderna and the student assistants for their invaluable help and profound commitment during the experiments.

Dr. C. Vorsatz for his helpful advices during the field trials and Dr. H. G. Schön for his expert advice during the statistical evaluation of the experiments.

Prof. Dr. E. Pawelzik and Dr. A. Werries, from the Department for Crop Science at the University of Göttingen; for their kind cooperation in implementing the detection of total phenolics.

Dr. A. Busch from the University of Osnabrück and Dipl. Ing. (FH) K. Mey from the University of Applied Sciences of Osnabrück; for their kind cooperation in implementing scanning electron micrographs.

The cooperating growers for providing their fields and greenhouses to conduct the experiments: Borrmann Gartenbau, Papenburg; Friedrich Schultz Gartenbau GmbH & Co. KG, Papenburg; Gemüsehof Andreas Wehmeyer, Herford; Gemüsehof Biewener KG, Melle; Mählmann Gemüsebau GmbH, Cappeln; Stefan Stegemeier Gemüsebau, Bielefeld.

The company Yara GmbH & Co. KG., Dülmen for providing the iodine fertilizer prototype.

The federal ministry for education and research (Bundesministerium für Bildung und Forschung, Förderprogramm "IngenieurNachwuchs", Grant No. 17N0210) and the University of Applied Sciences of Osnabrück for their financial support.